A Blueprint for a Safer Planet

*How to Manage Climate Change and Create
a New Era of Progress and Prosperity*

A Blueprint for a Safer Planet

How to Manage Climate Change and Create
a New Era of Progress and Prosperity

NICHOLAS STERN

THE BODLEY HEAD
LONDON

Published by The Bodley Head 2009

2 4 6 8 10 9 7 5 3 1

First published in Great Britain in 2009 by
The Bodley Head
Random House, 20 Vauxhall Bridge Road,
London SW1V 2SA

www.bodleyhead.co.uk

www.rbooks.co.uk

Addresses for companies within The Random House Group Limited can be found at:
www.randomhouse.co.uk/offices.htm

The Random House Group Limited Reg. No. 954009

A CIP catalogue record for this book
is available from the British Library

ISBN 9781847920379 (HBK)
ISBN 9781847920386 (TPB)

The Random House Group Limited supports The Forest Stewardship
Council (FSC), the leading international forest certification organisation. All our titles
that are printed on Greenpeace approved FSC certified paper carry the FSC logo.
Our paper procurement policy can be found at www.rbooks.co.uk/environment

Mixed Sources
Product group from well-managed
forests and other controlled sources
www.fsc.org   Cert no. TT-COC-2139
© 1996 Forest Stewardship Council

Printed and bound in Great Britain by
Clays Ltd, St Ives PLC

To Sue

Contents

Acknowledgements

The story of this book goes back to the summer of 2005 when Gordon Brown, then Chancellor of the Exchequer, asked me to lead a review on the economics of climate change. I reported to the prime minister, Tony Blair, and to Gordon Brown, and I am very grateful for their guidance and support both before and after publication of the Stern Review in October 2006, and for the fact that they encouraged me to be independent and to follow where the analysis took me. This work builds on the review, and all the acknowledgements to the many people who kindly worked on and helped with the review are also relevant here. In particular, I would like to recognise and thank again the team led by Siobhan Peters, and the special contribution of Dennis Anderson, who died in the spring of 2008.

This book arises from my experience of two years following the review, for much of which I have been working on the elements of a global deal and interacting with politicians around the world. I am very grateful to the political leaders I have met or worked with in this time, including President José Manuel Barroso of the European Commission, Prime Minister Manmohan Singh of India (a friend for nearly forty years), Chancellor Angela Merkel of Germany, Prime Minister Anders Rasmussen of Denmark, Prime Minister Kevin Rudd of Australia, President Nicolas Sarkozy of France and President S. B. Yudhoyono of Indonesia. Within the UK Cabinet, Douglas Alexander, Margaret Beckett, Hilary Benn, David Milliband and Ed Milliband have been constantly encouraging.

In the last two years many academic colleagues, both scientists and economists, have provided guidance and support. Among the scientists, I would like to thank particularly Myles Allen, Brian Hoskins, Jean Jouzel, David King, Robert May, Ernesto Muniz, Nicola Patmore, Vicky Pope, Martin Rees and John Schellnhuber; and among the economists, Tony Atkinson, Scott Barrett, Wilfred Beckerman, Francois Bourguignon, William Cline, Partha Dasgupta, Kemal Dervis, Nitin Desai, Peter Diamond, Ottmar

Edenhofer, Gunnar Eskelund, Ross Garnaut, Roger Guesnerie, Liu He,Geoffrey Heal, Dieter Helm, David Hendry, Claude Henry, Cameron Hepburn, Pan Jiahua, Paul Joskow, Paul Klemperer, James Mirrlees, Shandi Modi, David Newbery, Jyoti Parikh, Kirit Parikh, Amartya Sen, Achim Steiner, Laurence Summers, Laurence Tubiana and Martin Weitzman.

This book draws on a paper prepared in early 2008 and placed on the LSE website on 30 April 2008 called 'Key Elements of a Global Deal', and all the acknowledgements there are relevant here. I would like to mention in particular Eric Beinhocker, Nick Butler, John Llewellyn and Jeremy Oppenheim, with whom that very fruitful collaboration has continued. I am very grateful to McKinsey and HSBC for their collaboration and support on that paper. I advise the Group Chairman of HSBC, Stephen Green, on economic development and climate change. During that work and indeed over the last few years, I have benefited greatly from discussions with Adair Turner.

From the London School of Economics, I am very grateful for the great collegiality and counsel of Tim Besley, Alex Bowen, Howard Davies, Simon Dietz, Sam Fankhauser, Athar Hussain, Ruth Kattumuri, Danny Quah, Judith Rees and Lenny Smith.

There is a small group of friends and colleagues who have worked particularly closely on this book and I am deeply grateful to them: they are Claire Abeillé, Vicki Bakshi, Melinda Bohannon, Su-Lin Garbett-Shiels, Daniel Hawellek, Mattia Romani, Matthew Skellern, Chris Taylor, Tamsin Vernon, Dimitri Zenghelis and Tenke Zoltani. Eva Lee and Kerrie Quirk have provided tireless support, organisation and friendship and I owe them a great debt.

The later stages of this work have been aided by the very generous support of the Grantham Foundation and the substantial resources granted by Jeremy and Hannelore Grantham for the recently founded Grantham Research Institute for Climate Change and the Environment at the LSE. I am also grateful to Jeremy and Hannelore for their personal encouragement. The Economic and Social Research Council (ESRC) of the UK, through its new research centre on climate change at the LSE, has also supported this work. I continue to be grateful to the Suntory-Toyota International Centres for Economics and Related Disciplines at the LSE for providing my intellectual home, and to Leila Alberici, Sue Coles and Jane Dickson for their support.

I would like to thank my agent Andrew Wylie and his colleagues for their encouragement, guidance and support. The publishers Clive Priddle, of PublicAffairs Books, and Will Sulkin, of The Bodley Head, have provided helpful comments. Finally, special thanks are due to David Milner, who provided outstanding editorial guidance, and very wise, thoughtful and constructive crafting in the final stages of the book. It has benefited greatly from his collaboration.

Introduction

They were the right subjects. At the G8 Summit at Gleneagles in July 2005, the UK had proposed that discussions should focus on the two great challenges of the twenty-first century: fighting poverty, particularly in Africa, and combating climate change. Before the summit, and at the request of both the prime minister, Tony Blair, and Gordon Brown, the Chancellor of the Exchequer, I had led the writing of the Report of the Commission for Africa, published in March 2005, which was a key part of the preparation for the Gleneagles discussions. I attended the summit for the Africa part of the talks. Back in London a few days later, Gordon Brown and I asked the same question: why had we made so much more progress on Africa, in terms of explicit commitments, than on climate change?

Part of the answer was that there had been a very effective programme of global public pressure on poverty, including the campaign 'Make Poverty History' and the rock concerts Live 8, inspired by Bob Geldof. But we both felt there was something else. The analysis of evidence, options, policy and strategy was far more advanced on Africa, and there was strong African leadership of that analysis. In contrast, while scientists had done an excellent job in drawing attention to the very troubling magnitude of the risks of climate change, the policy and strategic options had not been laid on the table and examined rigorously in relation to the evidence. There was no shared understanding of what could work and why. This was the position from which the Stern Review on the economics of climate change was launched, and I set off on another major round of policy analysis, again reporting to both Tony Blair and Gordon Brown.

Working on the subject of climate change was very different from working on Africa. In 2005, as an engaged citizen, I was aware of the problems of climate change, but had not worked directly on them. I saw my task as providing an economic analysis firmly grounded in the principles of public policy and in the experience of making policy. What are the magnitudes, in economic terms, of the risks which the science has identified, and what are the policies and strategies that can help us manage those risks? These were the basic questions at the heart of the review. They required both a strongly international perspective, and the rigorous toolbox of the modern economics of public policy. That is what I hoped I could bring to the table.

I have worked on international issues and public policy in rich and in developing countries all my professional life. My first fieldwork at the end of the 1960s had been on tea grown on smallholdings in Africa, which profoundly influenced my thinking on public policy, as it showed what could happen if the entrepreneurship of the farmer was combined with the skills of the private tea factory in an environment shaped by sound policy, good infrastructure and effective agricultural extension. Of particular importance to me has been my work over more than three decades with splendid collaborators in the village of Palanpur in India. I have been following Palanpur and its economic and social change since I first went there in 1974–5 to study the effects of the Green Revolution, witnessing both the vulnerability and creativity of those in poor rural areas at first hand. Good policy, together with opportunity, unleashes entrepreneurship and achievement.

For twenty-five years, from 1969 until the end of 1993, I was an academic. After ten years' lecturing at Oxford, I had become a professor at Warwick and then at the London School of Economics, and also led an international life teaching and researching in India, China and Africa as well as at the École Polytechnique in Paris and the Massachusetts Institute of Technology in the USA.

Between 1993 and 2005, I had been more directly involved in the making of public policy as chief economist of the European Bank for Reconstruction and Development, chief economist and senior vice president of the World Bank, and then head of the Government Economic Service in the UK and Second Permanent Secretary at the Treasury, positions I occupied while writing the Report of the Commission for Africa. I am now an academic economist again, as I. G. Patel Professor of Economics and Government at the London School of Economics, chairman of the Grantham Research Institute on

Climate Change and the Environment, and director of the India Observatory.

When I began looking at the subject of climate change, what did I find? The first thing to hit me was the magnitude of the risks and the potentially devastating effects on the lives of people across the world. We were gambling the planet. I was also keenly aware of the importance of direct experience in mobilising action: unless people have seen or felt a problem, it is difficult to persuade them that a response is necessary. With the effects of climate change becoming apparent only over a long period of time, this is particularly challenging – and contrasts with my experience of working on Africa and India, and on development more generally, where the issues were often much more immediately visible.

My own experience of poor communities in Ethiopia, Kenya and India in the late 1960s and early 1970s convinced me as a young man to spend my life working on policy which could help create the circumstances in which people can alter their lives. That is why I am so concerned about climate change. Just as progress is starting to accelerate and we become more optimistic about battling poverty, we risk derailing that process through our inaction on climate change. And unless our action on climate change is consistent with rapid development, we will not get the global agreement that is fundamental for success. We will succeed or fail on these challenges together.

What is more, climate change is a problem which arises from a build-up of greenhouse gases over time and the effects come through with long lags of several decades. If the world waits before taking the problem seriously, until Bangladesh, the Netherlands and Florida are under water, it will be too late to back ourselves out of a huge hole. A special challenge of making policy here is that we are fast approaching a crisis which requires decision and action *now*, but we cannot yet directly experience the real magnitude of the dangers we are causing. And let us be clear, these dangers are on a scale that could cause not only disruption and hardship but mass migration and thus conflict on a global scale. They concern us all, rich and poor.

I was also troubled by some of the previous work on the economics of climate change precisely because it had not taken adequate account of the magnitude of the risks, and had not thought through sufficiently carefully the logic and ethics of policies which have long-term effects over several generations. By playing down the discussion of risks and of the ethics, it had missed or distorted issues which were absolutely central to policy. It had, in large measure, failed to apply precisely those techniques which were crucial to serious analyses of the problem: the economics of risk and the theory of public policy for the long term, where many existing markets, including capital markets, are distorted and hence of limited help in guiding policy. I saw an opportunity and a responsibility to change the economics of the discussion.

Another, much more positive thing to strike me, however, was that there is so much we can do to manage the risks and, if we act sensibly, to make a much better world. We can create a new era of progress and prosperity. We will discover new technologies and sources of energy along the way and make our energy supplies more secure. Ultimately, we will create a path of growth which can be sustained, our current high-carbon practices cannot deliver growth over the medium and long term, and generate major new opportunities for jobs and industry. But we have to act together, to create a global deal – this is a problem that is global in both its origins and its impacts.

That global deal must be *effective*, in that it cuts back emissions on the scale required; it must be *efficient*, in keeping costs down; and it must be *equitable* in relation to abilities and responsibilities, taking into account both the origins and impact of climate change. Climate change is an inequitable phenomenon: rich countries are responsible for most of the past emissions but developing countries are hit earliest and hardest. Climate change is likely to significantly disrupt their development. Looking forward, however, it will be the currently developing countries that will be responsible for most of the growth of emissions. Unless a deal is truly global it cannot work.

Difficult though it will be, I think such a deal can be constructed before the end of 2009. The pace of movement since the publication of the Stern Review in October 2006 has been remarkable. There have been so many important contributions to awareness including the splendid examples of Al Gore and the International Panel on Climate Change, who have worked for two decades to create an understanding of the science and the dangers, and who were very worthy winners of the Nobel Peace Prize in 2007. I would like to think that the Stern Review has also made a contribution in bringing the economics of policy to the table.

In his State of the Union message at the beginning of 2007, President Bush acknowledged for the first time that climate change was a serious challenge and closely linked to economic dependence on hydro-carbons. In the European Council of heads of EU governments in spring 2007, the EU committed itself to a 20% reduction in greenhouse gases by 2020, relative to 1990, and a 30% reduction in the context of a global deal. Just prior to the 2007 G8 Summit in Germany, China published its first climate change action plan. In the autumn of the same year, the Australian prime minister, John Howard, was ejected from office, partly on the grounds of his hostility to action on climate change, following devastating droughts. His successor, Kevin Rudd, immediately signed the Kyoto Protocol – thus depriving the United States of its one key ally in the 'coalition of the unwilling'.

The momentum for commitment to action has been sustained into 2008. In January, the European Commission set out detailed proposals for achieving the 20% cuts by 2020; the Indian government published its climate change action plan in June; at the Hokkaido G8 Summit, hosted by Japan in July, there was agreement to a 2050 target of a 50% cut in world emissions; in December, Brazil set strong targets for reducing deforestation – the source of around 70% of its greenhouse gas emissions – by 2020; Barack Obama has set a target of 80% reductions in the USA by 2050, spoke of a 'planet in peril' in his election-night speech and has made investment in 'green energy' a priority. There has been real progress in international understanding, and profound shifts in perspective.

In Bali in December 2007, the United Nations Framework Convention on Climate Change (UNFCCC) launched negotiations for reaching a deal in Copenhagen in December 2009 to take the place of the 1997 Kyoto Protocol, which runs out in 2012. The Copenhagen deal must be more ambitious, more international and much stronger. We must agree not only on our ambitions but on the details of action; the timetable is very tight. It will not be easy, but success is vital to the future of the planet.

If we do succeed, we will have created the potential not only to provide a serious response to the problem of climate change, but also to unleash an era of internationalism which could make the world much better at dealing with some of the other important international issues of our time, above all the fight against poverty. If we fail, the confidence and trust necessary to create and sustain an international agreement may be destroyed and the confidence of investors and markets, crucial for the real decisions that will make the necessary changes, will be undermined. Furthermore, addressing many of the obstacles to development, such as water availability, agricultural production, malaria and Aids, will become much harder and more costly. We have to see the issues of economic development and of climate change as parts of a whole.

We do, of course, have much to learn. While the science of climate change has been with us for nearly two hundred years, the study of policy in dealing with it is relatively young. There is a pressing urgency to settle policy now; we must act according to what we already know. Postponing a global deal will put both policy and markets in a limbo that could be very destructive. Those making the crucial investments in energy-related and other industries will not have the clear signals necessary to make considered and responsible decisions. The relentless logic of the flow of emissions adding to the concentrations of green-house gases in the atmosphere will place us in an ever more difficult position. We cannot afford to wait until we know everything with certainty. Indeed, the pervasive uncertainty makes it imperative that we act now to reduce the risks of a planetary disaster.

Some may argue that the turbulence in financial markets of 2007 and 2008 and the slowdown of the world economy should lead to the postponement of action on climate change. That would be a serious mistake. A key lesson of the present financial disarray should surely be that it is dangerous to ignore, or fail to recognise, the build-up of risk: this current economic crisis was fifteen or twenty years in the making. If we postpone action on climate change for fifteen or twenty years, our starting point will be much more difficult and risky. Further, in straitened times, energy efficiency will appear even more important. And we shall need a growth engine to take us out of the slowdown. Low-carbon energy could be a powerful driver: it is lasting and substantial; it is dynamic in its learning and global opportunities; and it is of real economic and social value.

It is true that some people, including politicians, have limited ability to concentrate and can only focus on one big issue at a time. But that is not true of all. And there may be enhanced sensitivity to cost increases, although these may be more easily managed as oil and gas prices fall in the slowdown. Leadership will be important, and in some countries, including the UK and the USA, we do see both a continuing concentration on climate change and recognition of its growth potential. It will be of crucial importance that public discussion and pressure for action continues.

A global deal will inevitably involve much detail and the economics of policy requires careful argument. I am not a campaigner; my purpose is to set out the analysis which explains why we must act, that we know how to act effectively and that the results will give us a better world for our children and grandchildren. I have tried to keep these arguments and analyses as simple as the subject allows. In my experience, if an economic or policy argument is sound it can usually be expressed in a way that is straightforward. But understanding policy analysis does require some application by the reader and it is important that the logic of arguments as well as the conclusion are understood. The reader will have to judge. There are some sections in the middle of the book which the non-economist might occasionally find a little difficult. I hope, however, that the basic thrust of the arguments, even in those sections, are apparent to all and that the non-economist reader who persists will find the details clear. Those who have an insatiable desire for the economics are provided with references which they can pursue.

The argument that the dangers are great and that the world should act strongly and urgently is, or should be, over. We should now be working on the policy and strategic response. The main purpose of this book is to argue that what we now understand is sufficient to point us unequivocally to measured and structured policies that involve deep cuts in emissions, in efficient and equitable ways, and that promote considered and careful adaptation to the effects of the climate change which will occur. I offer a blueprint of how to build a safer planet, or how to manage climate change while creating a new era of growth and prosperity. This is emphatically not a blueprint in the sense of a master plan of the kind that used to emerge from planning commissions in centrally planned economics. It examines what we now need: strategies, international understandings and policies that will guide action, correct the biggest market failure the world has seen and provide a framework for the entrepreneurship and discovery across the whole of business and society, which can show us how to achieve a cleaner, safer, more sustainable pattern of growth and development.

CHAPTER I

Why there is a problem and how we can deal with it

We start in a difficult place and we are following a dangerous path. At its most basic level, the science is simple and clear. Since the Industrial Revolution we have been emitting greenhouse gases at a faster rate each year than the planet can absorb, especially during the rapid and energy-intensive growth of the last sixty years. The gases trap the sun's heat as it is radiated back from the earth and cause global warming. This in turn causes climate change, with direct impacts on our livelihoods. Continuing with current practices will, by the end of this century, take us to a point where global warming in the subsequent decades of 5°C above pre-industrial times is more likely than not. Temperature increases of this magnitude will disrupt the climate and environment so severely that there will be massive movements of population, global conflict and severe dislocation and hardship.

The two greatest problems of our times – overcoming poverty in the developing world and combating climate change – are inextricably linked. Failure to tackle one will undermine efforts to deal with the other: ignoring climate change would result in an increasingly hostile environment for development and poverty reduction, but to try to deal with climate change by shackling growth and development would damage, probably fatally, the cooperation between developed and developing countries that is vital to success. Developing countries cannot 'put development on hold' while they reduce emissions and change technologies. Rich and poor countries have to work together to achieve low-carbon growth; but we can create this growth and it can be strong and sustained. And high-carbon growth will eventually destroy itself. We confuse the issues if we try to create an artificial 'horse race' between development and climate responsibility.

Scientists have been outstanding in marshalling the evidence on the risks. It is now the task of policy analysts and policymakers to construct strategies to reduce those risks and create a viable and attractive alternative to the high-carbon growth path we have been following. We can see the basic outlines of a way forward. But these policies and strategies are about incentives and opportunities for investments and technologies, and the relevant actions will mostly be from private investors, big and small. They must play a major role shaping policy.

The danger and the response

The danger from climate change lies not only, or even primarily, in heat. Most of the damage is from water, or the lack of it: storms, droughts, floods, rising sea levels. The levels of warming that we risk would be profoundly damaging for all countries of the world, rich and poor. A transformation of the physical geography of the world also changes the human geography: where we live, and how we live our lives.

We do not know for sure by how much global temperatures will rise if we follow 'business as usual'. I choose 5°C here to illustrate the risks because there seems to be around a fifty-fifty chance of eventual temperature increases around this magnitude if we continue along likely current growth paths, or follow business as usual for much of this century. Increases of 5°C would be devastating, but there are deeply worrying probabilities of being above 6°C or more next century. And even in a very optimistic assessment of the consequences of business as usual, we can still expect a rise of around 4°C, the effects of which would also be profoundly damaging, triggering unstable dynamics we do not yet fully understand. Indeed, such risks occur at lower temperatures and concentrations of greenhouse gases.

The seriousness of a 5°C increase is clear when we realise that in the last ice age, around 10,000 years ago, the planet was 5°C cooler than now. Most of Northern Europe, North America and corresponding latitudes were under hundreds of metres of ice, with human life concentrated much closer to the equator. We have to go back 30–50 million years, to the Eocene period, to find temperatures as high as 5°C above pre-industrial times. The world's land then consisted mainly of swampy forest. Temperature increases on this scale, and the consequent climate change, lead to massive dislocation, generate huge new vulnerabilities and redraw patterns of habitation. They cannot be understood in terms of the difference between Stockholm and Madrid, or Maine and Florida, or the idea that we might just need a little more air conditioning and flood defence.

The central message of this book is not, however, one of despair. These huge risks can be reduced drastically at reasonable cost, but only if we act together and follow clear and well-structured policies starting now. The cost of action is much lower than the cost of inaction – in other words, delay would become the anti-growth strategy.

The low-carbon world we must and can create will be much more attractive than business as usual. Not only will growth be sustained, it will be cleaner, safer, quieter and more biodiverse. We understand many of the necessary technologies and will create more; and we can design the economic, political and social structures that can take us there. We require clarity of analysis, commitment to action and collaboration.

It is neither economically necessary nor ethically responsible to stop or drastically slow growth to manage climate change. Without strong growth it will be extremely difficult for the poor people of the developing world to lift themselves out of poverty, and we should not respond to climate change by damaging their prospects. Moreover, politically it would be very hard to gain support for action by telling people that they have to choose between growth and climate responsibility. Not only would it be analytically unsound, it would also pose severe ethical difficulties and be so politically destructive as to fail as policy.

This is not to claim that the world can continue to grow indefinitely. It is not even clear what such a claim would involve; societies, living standards, ways of producing and consuming all develop and change. A picture of indefinite expansion is an implausible story of the future, but two things are key: first, to find a way of increasing living standards (including health, education and freedoms) so that world poverty can be overcome; and second, to discover ways of living that can be sustained over time, particularly in relation to the environment. Strong growth, of the right kind, will be both necessary and feasible for many decades.

The nature of the market failure

At the heart of economic policy must be the recognition that the emission of greenhouse gases is a market failure. When we emit greenhouse gases we damage the prospects for others and, unless appropriate policy is in place, we do not bear the costs of the damage. Markets then fail in the sense that their main coordinating mechanism – prices – give the wrong signals. That is, prices – of petrol or of aluminium produced with dirty energy, for example – do not reflect the true cost to society of producing and using those goods. In the language of economists, the social cost of production and consumption exceeds the private cost, so that markets without policy intervention will lead to too much of such goods being produced and consumed. By producing and consuming less of these products and more of others, we create economic gains that can make everyone better off. Markets with uncorrected failures lead to inefficiency and waste.

Market failures take many forms and much of economic policy is about correcting them. The most prominent are lack of information, abuse of market power and 'externalities'. An externality arises when the action of one person directly affects the prospects of another – discharging toxic waste into a river, building an eyesore and smoking in a restaurant, for example.

The appropriate response to a substantial market failure is not to abandon markets but to act directly to fix it, through taxes, other forms of price correction, or regulation. Acting in this way on climate change, with complementary policies on technology and deforestation, will allow continued and substantial growth and poverty reduction. Allowing the market failure to continue will damage the environment, curtail growth and lead to dislocation and conflict. To understand both the dangers and the appropriate policy requires an examination of how the market failure occurs and its effect on future generations.

Emissions are clearly an externality and are thus a market failure, but their impact is unlike that of, say, congestion or local pollution in four fundamental respects: the externality is long-term; it is global; it involves major uncertainties; and it is potentially of a huge scale. Greenhouse gas emissions constitute the greatest market failure the world has seen. Thus at the heart of economic analysis must be: the ethics of values both within and between generations; international collaboration; an appreciation of risk; and changes way beyond minor adjustments, or 'marginal increments' in the jargon so beloved of economists. Much of the analysis of climate change over the last two decades has been profoundly misleading because it paid no, or, at best, superficial, attention to some or all of these issues.

Our actions in the next thirty years, through investments, the generation and use of energy and electric power, the way we organise transport, and our treatment of forests, will determine whether or not we can keep climate change risks manageable. Less waste and more new technologies will be central to an effective response. The actions necessary to move onto a more sustainable pathway involve long-term planning: many of the relevant investments, such as power plants and buildings, have lifetimes of many decades.

In thinking through policy on climate change, and particularly its timing, we must also understand that there is a 'ratchet effect' of crucial importance. The flows of emissions build up into stocks or concentrations of greenhouse gases in the atmosphere that are very difficult to remove. Thus any delay means higher stocks and a more difficult and dangerous starting point for action. This ratchet effect together with long investment lifetimes, imply that the decisions, plans and incentive structures we make and create in the coming months and years will have a profound effect on the future of the planet.

In thinking about the long term how to reduce risks for future generations, ethical questions must be directly examined. All too often economists shy away from these issues, arguing that they are outside our subject, or say that ethics are 'revealed' by market behaviour and outcomes. Both arguments are mistaken. Economists provide analyses that inform political processes and policy and moral judgements, and that can help to shape questions. Economic analysis can show the implications of different sets of values for decisions and show inconsistencies. Moreover, while markets can provide some limited information relevant to values, there is no way that they can settle debates over which values should be used to guide decisions of this magnitude, collective responsibility and timescale.

The second special feature of the externality is its global nature. Greenhouse gases have the same effect on global warming whether emitted from London, New York, Sydney, Beijing, Delhi or

Johannesburg; hurricanes and storms hit New Orleans and Mumbai; flooding occurs in England and Mozambique; droughts in Australia and Darfur; sea-level rise will affect Florida and Bangladesh.

There is a double inequity here: the poor countries are least responsible for the existing stock of greenhouse gases, yet they are hit earliest and hardest by climate change. At the same time, the rapid growth of China, India and others is already making them important emitters – China has overtaken the USA to become the world's largest producer of greenhouse gases. Indonesia and Brazil are the third and fourth largest emitters, mainly as a result of deforestation and peat fires.

The rich countries have major historical and other responsibilities, and must show leadership. Without it, global action will fail. But the future of the climate will largely be shaped by the developing countries: in population terms, it is their planet. Already, the rich countries constitute less than one in six of total global population; by 2050, they will be only one in nine. The large developing countries will be central to the design and execution of international action to protect their future. It is profoundly inequitable that the difficult starting point is largely as a result of actions by the developed nations, but the numbers on population and future emissions are such that a credible response cannot come from the rich countries alone.

The third and fourth special features of the externality, the centrality of risk and the scale of possible damage, shape both the structure of the argument and the method of analysis. This is not an investment project like a new road or a bridge. The costs and benefits of such projects can reasonably be understood in terms of a marginal change, set in the context of a *given* growth path for the entire economy. What we are discussing with climate change are strategies concerning patterns of growth, or possible decline, for the world economy as a whole in the context of uncertain outcomes. The analytical tools and policy constructs must be capable of taking on these issues directly. All too many discussions – and it is astonishing that they have done so – see policy on climate change as a single-investment decision, analogous to a new bridge. Standard, marginal cost-benefit analysis is appropriate for the latter kind of decision. For climate change, however, the relevant economics are much more difficult and profound.

Shaping policy

The central economic criteria in forming policy must be: effectiveness in reducing emissions on the scale required; efficiency, to keep costs down; and equity, in recognising differences in incomes, technologies and historical responsibility. The earlier we start to put the policies in place, the longer we have for a calm and measured response. Delay now and haste later not only build up damage but also risk expensive mistakes in investment decisions. The greater the coordinated involvement of all emitters, the more successful, cheaper and equitable are the actions and outcomes.

Good policy can avoid many of the more devastating scenarios that could arise in the second half of this century and into the next. Throughout the world we are seeing record events, whether they be floods, droughts or hurricanes. We cannot directly attribute any one of them to climate change, but we can say that the frequency of similarly severe events will rise. We are already seeing the effects of a 0.8°C temperature increase; we have around 1.5°C or 2°C more to come, even if we act strongly and sensibly immediately. The costs of adapting to these changes will be very large; planning for adaptation to climate change should start now in all countries of the world.

The challenges for developing countries will be particularly severe. To ignore a changing climate and what it implies would simply make for bad investments. Adaptation must be part of development. At the same time, development itself will be very important in adapting, as it encourages economic diversification and a more flexible workforce, both of which reduce vulnerability. It also generates the income necessary for robust investment and it fosters greater technical knowledge. In all these fundamental ways, development is the most important form of adaptation.

All countries will have to reorientate their economies, both to strengthen resilience to the inevitable effects of climate change and to lessen the use of carbon in order to reduce substantially those risks in the future. This is a story about the economic management of investment and growth from the perspective of both adaptation and mitigation.

Leadership at the most senior level is vital, as that is the only place at which the kind of mutual understandings and trade-offs that are necessary for a global deal can be made. Collaboration may involve, and be helped by, understanding on other issues such as trade, health and financial stability. Furthermore, the process of reaching a global deal may lay the foundations for future international cooperation on broader issues. This wider context means that the global deal cannot be left *only* to environment ministries, important though they are, and however strong and able they might be – *all* ministries must be involved and committed, and the deal must be in the hands of heads of government. Only if world leaders give this issue the attention and priority it deserves can a deal be made which reflects the magnitude of the risks and the scale of action required.

The logic of the problem structures the argument of this book. The first half sets out the challenges in terms of the risks of climate change and the scale and nature of action, from a global perspective, necessary to manage those risks. The second half analyses the policies required to promote action, proposes the basics of a global deal and examines the challenges of how to build and sustain that deal.

CHAPTER 2

The dangers

The scale of risk and uncertainty

In broad and basic terms, the process of climate change starts with the actions of people and ends with the impacts on people. First, we cause, via our activities, emissions. Second, because the planet cannot absorb them all, we add, year by year, to the concentrations or stocks of greenhouse gases in the atmosphere – the processes of absorption and addition constitute the 'carbon cycle'. Third, energy is then trapped in the atmosphere, causing warming – climate scientists use the term 'climate sensitivity' to describe the amount of warming arising from a given increase in concentrations. Fourth, warming causes climate and environmental change which, in the fifth and final link, ultimately affects our lives and livelihoods.

Each of these links involves substantial risk and uncertainty which, cumulatively, are very large, not only because our knowledge is incomplete but also because solar, planetary and other processes have an inherent randomness. The whole subject of policy on climate change involves decision-making under risk and uncertainty.

The links involved in the chain also involve time lags of varying lengths, some of which are much greater than others. The lags between the first two stages – from people to emissions, and from emissions to increased stocks – are quite short. The lags between the third and fourth – from increased concentrations to warming, and from warming to climate and environmental change – can be quite long: there might be decades between increases in stocks and temperature changes. Some of the effects of temperature increases, such as rising sea levels, can take centuries to appear. These lags make it much more difficult to agree on the need for action to rein in emissions, since unlike, say, having someone smoke a cigar next to you in a restaurant, there is no immediate reminder of the negative effects. Yet if action is delayed, concentrations build up, and by the time the consequences are apparent, the conditions for further temperature increases will already have been created. From that point, it will also take a long time to reduce concentrations. On the other hand, the lags allow us to look ahead and plan how to adapt.

Notwithstanding the uncertainties and the lags, the science of the warming induced by emissions is clear and of long standing. In the 1820s, the French mathematician and physicist Joseph Fourier examined the heat balance of the earth from the perspective of incoming solar radiation and outgoing infrared radiation, concluding that as the planet was around 30°C warmer than he expected, something was trapping infrared radiation.[1] Thirty years later, the Irish physicist John Tyndall identified the molecules (including CO_2 and water vapour) that were blocking the radiation. Those molecules became known as greenhouse gases.

At the end of the nineteenth century, the Swedish chemist Svante Arrhenius was the first to present calculations on the temperature increase that might result from a doubling of CO_2 concentrations in the atmosphere from their mid-nineteenth-century levels of around 285 parts per million (ppm).[2] His contribution was to point out quantitatively the relevance of greenhouse gases for the global climate. He was not in a position to take into account various interactions and 'feedbacks', including in particular the effect of water vapour, that modern approaches now recognise. Recent simulations suggest that a doubling of concentrations would cause an equilibrium or eventual increase in average global temperature of between 1.5°C and 5°C. The Hadley Centre in Exeter, a branch of the UK's Meteorological Office and one of the leading centres of climate modelling research in the world, has estimated that the likely increase is centred around 3.5°C above mid-nineteenth-century levels.[3]

Modern climate science, drawing on much more evidence and on computing power that allows greater regional resolution concerning the role of oceans, ice caps and so on, gives increasingly detailed predictions about many aspects of the world's climate and geography, including ocean temperatures and currents, as well as atmospheric temperature. This means that, when different models with divergent explanations of temperature increase are compared, we must also compare the implications for other features of the earth. Most attempts to explain observed temperature increases without including human-induced greenhouse effects fail.

We are now able to analyse and model how concentrations might grow by looking at the processes of emissions of greenhouse gases and their absorption, and to examine in much more detail the possible consequences of those increases. Also, each link in the chain can be analysed in terms of the risks and uncertainties involved.

Before looking at each of the five links in the chain, it is important to say that I generally use the words 'risk' and 'uncertainty' interchangeably in this book, but sometimes the distinction needs to be made because it affects the thinking about policy. As first suggested by Frank Knight of the University of Chicago in the early 1920s, a situation involves 'risk' if we have evidence on, or are able to guess at, the probability that each outcome will come to pass; 'uncertainty' is the term given to circumstances where we have some idea of possible outcomes, but are not in a position to provide educated guesses at the probabilities. As climate science advances we are able to move from the analysis of uncertainty towards the analysis of risk. Until fairly recently, scientists would have been able to indicate a range of possible temperature increases from a given concentration level, but would have found it very difficult to attach probabilities to different parts of the range. In other words, they would have been able to identify the dangers, but would have been unable to provide much detail on how likely they are.

Over the last five or ten years, however, scientists have been able to gain some understanding of the probability of different temperature increases and other climate outcomes by examining many models, or versions of models, under different assumptions. They then have a large number of examples of possible outcomes arising from, for instance, a given increase in concentrations of greenhouse gases. There is then a collection, or 'ensemble', of cases to look at. If we thought that, given current knowledge, all of the examples in the ensemble were equally likely, then the probability of being above, say, 5°C at some point in the future, is estimated via the fraction of the cases in the ensemble that turn out to be above 5°C at that time. More generally, assessing the likelihood of each of the different examples gives an estimate of overall probabilities.

This increasing knowledge does not mean, however, that we can now estimate all the relevant probabilities. Major uncertainties, in the Knightian sense, remain. For example, we still have much to learn about the way in which oceans and their absorptive capacities are likely to change. We must consider decision-making under uncertainty as well as decision-making under risk. I generally concentrate on risk, for it is here that principles, theory and common sense are better developed, and the science has now put us in a position to apply these principles systematically.

From people to emissions

The rate of growth of economic activity in the industrialising parts of the world accelerated dramatically from the mid nineteenth century onwards, while the form of that activity (the rise of industry and the relative decline of agriculture, for example) became much more hydrocarbon-intensive. These three effects – growth, industrialisation and hydrocarbon use – combined to increase greenhouse gas emissions.

The second half of the twentieth century saw a sharp increase in rates of growth of emissions as the world recovered from the Great Depression and the Second World War, and more countries industrialised. The growth of China accelerated after the end of the Cultural Revolution in 1976 and the reforms initiated by Deng Xiaoping, which began in 1979. China's growth rates, averaging about 8% over the last three decades, have led to a doubling of output roughly every nine years. The growth of energy consumption – mostly hydrocarbon-based, with a particular focus on coal for electricity generation – has been on a similar scale. China is now the world's largest emitter of greenhouse gases, with total emissions a little higher than the USA. That said, its population is four times larger. For the twenty largest emitters, the pattern of emissions in 2004 is shown below in terms of CO_2 equivalent[4] (CO_2e):

Emissions by country

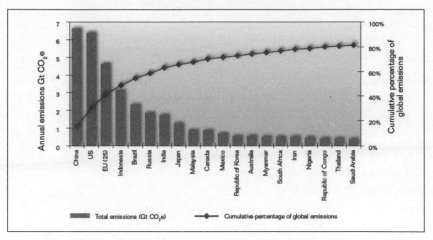

Source: Garnaut (2008), UNFCCC (2007) 2004 data for US, EU (25), Russia, Japan and Canada; Department of Climate Change (2008) 2004 data for Australia (using UNFCCC accounting); and World Resources Institute (2008) for other countries (2000 data except for CO_2 emissions from fossil fuels, which is for 2004).

To look at the time trends at the country level, however, it is necessary to switch to CO_2 only because figures for CO_2e are patchy. Average per capita CO_2 emissions across the world have increased around 15% in the last three decades. From the graph below, three different trends are evident: per capita emissions in high-income[5] countries, with a population of 1.2 billion in 2007, have been stable or slightly decreasing, mostly due to deindustrialisation of their economies; per capita emissions in developing countries, with a population of around 5.2 billion, have increased significantly (from 1 tonne to 4 tonnes per annum), particularly in recent years; per capita emissions from the least developed countries, with a population of around 0.7 billion (a subset of the 5.2 billion) have been stable and very low (around 0.2 tonnes per annum). The world population increase in those thirty years was around 50%, with the result that total CO_2 emissions increased by around two-thirds.

These same features point us to likely developments under business as usual. Key drivers of growth in overall emissions, CO_2e will be per capita growth of emissions in the developing countries plus population growth in those countries of 30–40% in the next forty years. A little arithmetic can illustrate. In 2008, total developing countries' emissions

CO_2 emissions per capita (1976–2004)[6]

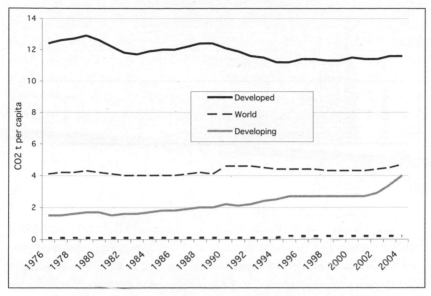

Source: Climate Analysis Indicators Tool (CAIT) Version 5.0 (Washington, DC: World Resources Institute, 2008).

and total developed countries' emissions were probably around 30 Gt CO_2e and 20 Gt CO_2e respectively. Developing countries will have a population of around 8 billion by 2050. If the per capita CO_2e emissions of developing countries increased from around 5 tonnes now to about 10 tonnes under business as usual (a modest assumption), then their total CO_2e emissions would grow to approximately 80 Gt. If the total for rich countries stayed around 20 Gt, then the total annual emissions in 2050 would be around 100 Gt CO_2e.

The example is for illustration only. One could tell a business as usual story with much higher growth in emissions, since incomes in developing countries as a whole may go up by a factor of five or more between now and 2050 (and see below on emissions per unit of output). Or one could tell a business as usual story with somewhat lower growth in emissions because there is strong substitution away from hydrocarbons as prices rise. It is important to recognise the range of possibilities and the key assumptions. It should be clear, however, that there is a real possibility of emissions reaching very high levels if developing countries are not part of an international effort to restrain them.

China's growth has lifted 300–400 million people out of severe poverty (roughly measured as the equivalent of $1 per person per day)[7] since 1979. It is the most remarkable reduction in poverty in the history of the world, far exceeding in scale anything witnessed by the West. Of course, this growth has been accompanied by severe stresses, such as rapidly increasing inequality and a deteriorating environment, but these should not blind us to the great rewards in the form of better living standards for many. The rest of the world has, in turn, also bene-fited from cheap Chinese-manufactured exports and China's increasing demand for imports.

The location of production has changed, and many have gained from the new division of labour. Those who consume and those who produce together determine the patterns of location of production and of emissions. If country A produces financial and other services and buys manufactures from country B, they are jointly shaping not only the patterns and levels of production but also of emissions. Thus one must take care in ascribing responsibility for emissions in a simple-minded way solely to the country in which production and emissions take place.

Other big countries have seen rapid growth in the last twenty years. Vietnam, with a population of 86 million, has had economic growth rates of around 7% per annum in that time, and India has had rising rates of growth for the last quarter of a century, particularly since the economic reforms of the early 1990s. In the last five years, India has seen growth of around 8%, yet per capita emissions are still below 2 tonnes CO_2e per annum. Sub-Saharan Africa has grown more rapidly in the last ten years than in the preceding few decades with annual growth of around 5%, although its per capita emissions are still mostly less than 1 tonne CO_2e per annum and in many cases a small fraction of 1 tonne.

The relationship between emissions and output, though positive, is not clear cut. As a country becomes richer, a greater proportion of economic activity is devoted to service industries, which are generally less emissions-intensive than manufacturing or extractive industries. But for poorer countries the share of manufacturing also rises as that of agriculture declines. Attitudes to the environment change, as does people's willingness to support more environmentally friendly tech-nologies. Natural resource endowments and industrial efficiency will also affect the relationship between output and emissions, as will policy.

Over the last thirty years, most economies in the world have been using less and less energy to produce a unit of income. The gain in energy efficiency is most striking in the case of China: on average the production of one unit of Chinese national income requires eight times less energy today than in 1980. While this progress is remarkable, a unit of Chinese income still requires twice the quantity of energy necessary to produce a comparable unit of income in Europe, mostly reflecting the different sector mix of the two economies.

The CO_2 content of the average unit of energy consumed shows a sharp reduction for many richer countries over the last fifty years. The trend is very different for low-income countries. Particularly in India and China, the CO_2 content of the average unit of energy has increased. The trend in CO_2 emissions per unit of income is slightly downward for industrialised countries. China's CO_2 emissions per unit of income was also trending downward over the period 1980–2000, as China decreased substantially the energy content of the average unit of income. But it then increases sharply over the last few years: this is the combined effect of the flattening in the trend of decreasing energy consumption per unit of income, and the sharp increase in the CO_2 content of energy.

Ultimately, the message is that it is the combination of energy intensity of the economy and the carbon content of energy that will matter in terms of the CO_2 footprint of income, and that, at least among quickly industrialising countries, the former has been falling, but the latter has not. If income is rising quickly and the CO_2 content of income is rising, then emissions will rise even faster than income.

Since the mid nineteenth century, cumulative emissions have been about 1,200 gigatonnes CO_2e, and concentrations have grown from 285 ppm to 430 ppm CO_2e today.[8] Approximately 70% of all emissions from 1850 to 2000 took place in the second half of the twentieth century, and stocks increased from 330 ppm to 410 ppm.[9]

Currently, developed countries – which contain around 1 billion of the 6.7 billion people alive today – are the source of around 70% of the emissions since 1950.[10] In the future, countries now considered 'developing' will be the source of the bulk of emissions.[11]

If developing countries see emissions continue to increase at their present annual level of 3% or 4%, then in twenty years by themselves they will emit as much as the current total world emissions of over 50 Gt CO_2e per annum, which by then would constitute more than 70% of global emissions (assuming developed countries' emissions are moderately flat under business as usual for the next twenty years). Unless developing countries play a strong role in a global deal to combat climate change in the next two decades, it will be impossible to cut emissions on anything like the scale required.

Many studies after 2000, including some of the modelling in the Stern Review, relied on the Special Report on Emissions Scenarios (SRES) of that year, developed by the Intergovernmental Panel on Climate Change (IPCC).[12] These scenarios are in urgent need of revision; the Australian economist Ross Garnaut has made an important contribution to this process, as has the International Energy Agency (IEA), although the IEA is under strong political pressure to moderate its estimates of increases in particular countries. Contrary to claims by some that the Stern Review exaggerated the likely growth in emissions under business as usual by picking one of the higher SRES paths, all the evidence that has since accumulated suggests that emissions are likely to be considerably greater than the highest SRES scenario. For example, Australia's Garnaut Climate Change Review suggests a forecast that in the next twenty-five years, China is likely to produce emissions similar to those of the USA over the last century.[13] This is not to criticise China, the rapid growth of which has, as already mentioned, brought many benefits to its own people and to the world that are entirely to be welcomed. And, contrary to suggestions by some that China does not take climate change seriously, it is actively exploring ways to reduce energy consumption and emissions.

The challenge is to achieve low-carbon growth for the world as a whole. The vital point here is that recent analysis of emissions under business as usual indicates that they are growing substantially faster than previously thought. This makes the danger of doing nothing all the more worrying.

From emissions to stocks

The second link in the climate change chain is the contribution of greenhouse gases to stocks in the atmosphere. How much is added depends on how much of the emissions are absorbed, which is determined by the absorptive capacity of key aspects of the planet, particularly oceans and forests. The capacity of the oceans now looks to be lower than earlier models had indicated – hence we can expect a larger increase in stocks for a given amount of emissions.

The absorptive capacity of forests varies greatly with type and location. Much more CO_2 is stored in native tropical forests (approximately 500 tonnes per hectare) as opposed to northern native forests (perhaps a tenth of that or less). Annual deforestation rates for native tropical forests between 2000 and 2005 are around 11% in Nigeria and Vietnam, 2.5% in Indonesia and 1% in Brazil.[14] Given the very large forest areas in Brazil and Indonesia, more than half the emissions from deforestation are from those countries and the average annual rate of deforestation for the world for tropical forests is around 2%. It is estimated that the Amazon rainforest as a whole stores about ten times as much carbon as is currently emitted globally per year. Its potential collapse as temperatures rise by 2 or 3°C would lead to substantial additional emissions of CO_2.

The combination of rising emissions and decreasing absorptive capacity has implied that, not only has the concentration of greenhouse gases been going up, it has also been doing so at an increasing rate. From 1930 to 1950, the concentration of all Kyoto gases[15] increased by about 0.5 ppm per annum, from 1950 to 1970 by around 1 ppm per annum, and from then until 1990 the rate of increase doubled again. In the past decade it has been around 2.5 ppm a year. This gives some indication of the scale of the problem and the dangers of delay.

Strong world growth is likely to continue, particularly with the rapid growth of developing countries. Reasonable emissions projections suggest that, under business as usual, the annual augmentation to stocks over the first half of this century might be 3–4 ppm, with more than this in the second half. This would take levels from around 430 ppm now to 580–630 ppm by mid century, and to around 800–900 ppm by the end of the century.[16] Add possible temperature increases from methane released from thawing permafrost and more rapid loss of rainforests, for example, and the numbers could be still higher.

From stocks to rising temperature

Scientists often use the term 'climate sensitivity', in a very specific sense, to describe the eventual temperature increase which might follow from a doubling of concentrations from the 1850 level of around 285 ppm CO_2e.[17] Given the uncertainties, climate sensitivity is described not by a single temperature increase but as a range, with estimated probabilities for different parts of the range. It must be remembered that these temperatures represent averages across the whole surface of the planet, both oceans and land, and over different times of year and weather patterns. The detailed variation will, of course, be of great importance to livelihoods and activities.

Concentration levels and temperature increases

Stabilisation Level (in ppm CO2e)	2°C	3°C	4°C	5°C	6°C	7°C
450	78	18	3	1	0	0
500	96	44	11	3	1	0
550	99	69	24	7	2	1
650	100	94	58	24	9	4
750	100	99	82	47	22	9

Likelihood, in percentages, of exceeding a temperature increase at equilibrium, relative to 1850. Source: Stern Review, Box 8.1, p. 220, with some added information.

These figures are taken from the Hadley Centre models, which are generally somewhere in the middle of the range of climate models – neither overly gloomy nor overly sanguine. Other authors have demonstrated that current models have a much wider range of climate sensitivities under very plausible assumptions.[18]

A cautious estimate of the consequences of business as usual might be a concentration level of, say, 750 ppm CO_2e by the end of this century. Even if concentrations were held at this level, the implications would be a roughly 50% chance of temperature increases in the first half of the next century of above 5°C relative to 1850. For the 850 ppm which is perhaps more likely, the probability of being above 5°C would be around 70%, with around a 40% chance of being above 6°C. These are not small probabilities of big changes; they are large probabilities of enormous changes.

We should also be clear that even if concentrations were held below 500–550 ppm, the probabilities of a 2–3°C increase are high. Given that we are already at 430 ppm, we have probably missed the chance of keeping emissions below 450 ppm. Limiting temperature increases, with high probability, to 2°C – which is often advocated on the grounds that anything higher would be dangerous – is a goal that is unlikely to be achievable, unless we discover and implement ways of extracting greenhouse gases from the atmosphere on a large scale. At 500 ppm CO_2e, the chances of exceeding 2°C are over 95% – but we would still have a strong chance of staying below 3°C. Many scientists and environmentalists have argued for a goal of no more than 400 ppm CO_2e (roughly 350 ppm CO_2) on the grounds that these are the concentrations within which species, including humans, evolved. The only way 400 ppm can be achieved (or indeed 450, which we will reach within ten years) is by not only stopping concentrations from increasing but also eventually reducing them. Given the long lifetime of CO_2 and the difficulty of extracting it, the necessary reduction might take hundreds of years during which strong warming will occur. Nevertheless, the risks involved in 500 ppm or even 450 suggest that we should be thinking of bringing down concentrations below 500 in the very long term.[19]

From rising temperature to climate change to the human impact

An increase in global mean temperature has strong impacts on key aspects of our physical environment, including the sea level, precipitation rates, hurricane intensity and frequencies and so on. This link, from temperature to climate change, is the fourth in the chain – which starts with people and their emissions and ends with the impact of climate change on people. It is a complex process with various feedback effects involving, for instance, the amount of water vapour in the atmosphere or the sea ice reflectivity. One characteristic feature of climate change is higher variability on key dimensions. Record hot summers can be followed by cold spells, heavy monsoon rainfalls by intense dry periods. This in turn affects us humans – the fifth and final link in the chain. It now appears likely that these two links involve greater risks, and more severe and rapid effects, than were assumed in the Stern Review and in the Fourth Assessment Report of the IPCC.

We must recognise that even if we act responsibly it is quite likely that we will see temperature changing rapidly, in historical terms, to around 2–3°C higher than in 1850. What would a world with these types of temperature increases look like?

Looking back, we have evidence from direct temperature increases covering 150 years. For an understanding of climate change we have to look back much longer than that. Ice cores provide evidence on both CO_2 and temperature. These can take us back 700,000 to 1 million years – although the further we go back, the bigger the ranges of uncertainty. These estimates of CO_2 and temperature can then be related to corresponding archaeological and biological evidence. To go back still further we have to rely largely on palaeo-climatology. Levels of CO_2 can be deduced from, for example, fossils found in chalk, and temperature levels from organic material, such as fossilised leaves. These two can then be related to corresponding archaeological and biological evidence.

A key aspect of climate change is the sea-level rise. Given the complexity of the deep ocean systems, it is notoriously difficult to forecast the change in sea level associated with climate change. Consequently, substantial differences between scientific studies exist and the uncertainty around best estimates is again large. The fourth IPCC report projects a range of 18–59 cm over the next century. Since oceans react very slowly, the eventual rise might be very much larger and estimates go as high as several metres per degree centigrade of warming. Further to this, the potential disintegration of the Greenland and West Antarctic ice sheets could constitute important – and not very well understood – tipping points.

What would a sea-level rise of just one metre imply for the different regions on the planet? By far the most affected is Asia: it has the longest exposed coastline, the largest number of people threatened and the highest value of assets at risk. Based on today's situation, around 150 million people in Asia and $1 trillion of economic assets would be directly exposed. Given the fast population growth and the enormous expansion of port cities these figures will be much higher by the time a one-metre sea-level rise would have occurred. While in 2070 the exposed number of people will be more than 10 million in cities such as Kolkata, Mumbai and Dhaka, the potential monetary value of economic damage would be higher in places like Guangzhou in Guangdong Province of China, New York and Miami. Despite the fact that flood defences in Europe are comparatively solid, the potential economic damage to some of the major port cities like Rotterdam or Hamburg is very large. In some regions adaptation will not be feasible either for economic or geographical reasons. And remember, the eventual sea-level rise from just 1°C might well be several metres, not just one metre, and while sea-level rise takes a long time to appear, the process once started is extremely difficult to control.

With a temperature increase of 2–3°C many parts of the world will experience severe dislocation, with rising sea levels, a greater frequency of intense storms and hurricanes, the melting of glaciers and snows which will lead to torrents and flooding in wet seasons, droughts in many parts of the world, and a high risk that the major rainforests will collapse. In the last two or three decades we have already seen some of the first forerunners of what might happen, and we are 'only' around 0.8°C above 1850 now.

Sub-Saharan Africa saw the warmest and driest decade on record between 1985 and 1995. This heating trend will continue under global warming and lead to droughts, fires and stress on agricultural productivity.[20] One of the early warnings was extensive fires along the coast in the Western Cape Province in South Africa fuelled by record temperatures of 40°C in January 2000. Water reservoirs will continue to decrease. The surface area of Lake Chad has decreased from 25,000 square kilometres in 1963 to 1,350 today. Modelling studies indicate the severe reduction results from a combination of reduced rainfall and increased demand for water for agricultural irrigation and other human needs.[21]

The Himalaya region and the areas fed by its waters are likely to experience unprecedented flooding as glaciers melt and glacial lakes swell. Average glacial retreat in Bhutan for instance is 30–40 metres per year. Temperatures in the high Himalayas have risen by about 1°C since the mid 1970s.[22] Ice-core records from the Dasuopu Glacier indicate that both the last decade and the last half-century have been the warmest in 1,000 years.[23] While melting glaciers lead to flooding and landslides in the medium term, water shortages are the consequence in the long term. The Himalayan glaciers supply more than 8 million cubic metres every year to Asian rivers, including the Yangtze and Yellow in China, the Ganges, Brahmaputra and Yamuna of India and Bangladesh, and the Indus in Pakistan.[24] In the Ganges itself, the loss of glacier meltwater would reduce July–September flows by two-thirds, causing severe water shortages for 500 million people and 37% of India's irrigated land.[25]

While coastal areas in North America will be affected by sea-level rise and increased hurricane intensity, the impacts of climate change on the mainland are uncertain. Observations in the recent past suggest that extreme events might be more important for this region than sea-level changes. A striking example was 1998: the Black Hills in South Dakota received 260 cm of snow within five days in February, almost twice as much as the previous single-storm record for the state.[26] The same year an unprecedented autumn heatwave from mid November to early December broke or tied more than seven hundred temperature records for daily-highs from the Rockies to the East Coast. Temperatures were over 20°C as far north as South Dakota and Maine.[27]

Europe is already seeing strong impacts of climate change from global temperature increases of less than 1°C. In the heatwaves of 2003, 35,000 more people than usual died. Southern Europe is experiencing severe water shortages. Threats of flooding in London have risen sharply. What we are seeing in extreme circumstances over the last decade is likely to be average by mid-century, with still greater extremes relative to those averages.

In terms of human disaster, the poorest countries will be suffering most, given their limited capacity to respond to the challenges. More than 300 million people are currently exposed to tropical cyclones which are expected to become more intense. Hundreds of millions of people might additionally be infected by malaria and a similar figure holds for those suffering from undernourishment caused by droughts and rural poverty in addition to current levels. Bangladesh, a country with some 150 million people and an average income of $1,300 (measured in purchasing power parity), is already experiencing regular flooding.

Although in the short term some areas are likely to benefit in terms of less severe winters and improved opportunities for agriculture or sea routes, all countries will have major challenges in adapting to rapid climate change. Those that experience some offsetting benefits in the short term will also have major short- to medium-term adjustment costs. And those benefits will themselves be transient if business as usual continues.

If we do not act responsibly, it is likely that by the end of this century or some time in the first half of the next century, we will see temperature increases of 4–5°C or higher relative to 1850. It is very difficult to describe such a world; the science has much less to go on.

But it would, in all likelihood, be a radical transformation of the world we know.

The last time the world was 4–5°C above where we are now was 30–50 million years ago, in Eocene times, when much of the planet was swampy forest and there were alligators near the North Pole. The existence of alligators in that part of the world might not be a big issue per se – the point is that the location of many species, including humans, would be radically different and many would not survive. Some areas, probably much of Southern Europe, might become deserts. Most of Florida and Bangladesh would eventually be submerged. Of great importance here is that the pace at which temperature changes occur would be extraordinarily rapid in relation to historical and evolutionary time. Increases of the scale of 5°C could happen, indeed are likely to happen, within the space of a century or two if we do not act.

An indication of the redrawing of the planet that temperature change on this scale could involve can also be obtained from a 'recent' climate change in the other direction: the last ice age, around 10,000 years ago, when temperatures were 4–5°C lower. Ice sheets came down to latitudes just north of London and just south of New York. Human beings were concentrated much closer to the equator than now. When the ice melted, the UK separated from Continental Europe. I do not want to debate the pros and cons of that separation, although it makes for enjoyable dinner-party conversation.

The point is that with temperature changes of this magnitude, the physical geography is rewritten. If the physical geography is rewritten then so too is the human geography of the world. There would be movement of people on an immense scale. The lessons of the last few hundred years surely tell us that the movements of billions of people in a fairly short period of time would plunge the world into massive and extended conflict.

The deniers

How is it that, in the face of overwhelming scientific logic and evidence, there are still some who would deny the dangers and the urgency of action? Not surprisingly, the loudest voices are not scientific, and it is remarkable how many economists, lawyers, journalists and politicians set themselves up as experts on the science. It is absolutely right that those who discuss policy should interrogate the science because the implications for action are radical; however, they should also take the scientific evidence seriously and recognise the limitations on their own abilities to assess the science.

Contrary to the narrative that some have tried to impose on the debate, climate change is not a theory struggling to maintain itself in the face of problematic evidence. The opposite is true: as new information comes in it reinforces our understanding across a whole spectrum of indicators. The subject is full of uncertainty, but there is no serious doubt that emissions are growing as a result of human activity and that more greenhouse gases will lead to further warming.

The last twenty years have seen special and focused attention from the IPCC, which has now published four assessments, the most recent in 2007. With each new report, the evidence on the strength and source of the effects, and the magnitude of the implications and risks, has become stronger. Some people accuse the IPCC of having institutional and procedural structures which predispose it to alarmism. In fact, the IPCC is structurally conservative and requires very tight consensus among scientists from many backgrounds and nationalities. As a result, statements are muted and it is likely that risks are understated. It mostly confines attention to the period until 2100, when the lags are such that still bigger damages appear later; and it leaves out effects which are likely to be important but on which strong, detailed quantitative evidence has yet to accumulate sufficiently.

Some of the marginally more sophisticated attempts at obfuscation focus only on mean expected temperature increases in the short term, rather than looking at a longer horizon or at the very real possibility of much higher increases. Look, they argue, the IPCC does not expect a temperature increase of much more than 2.5–3°C by the end of the century; we can cope with that. This is a classic example of the misuse of evidence to divert attention from the main point – how to control the risk of bigger increases. By focusing on the limited time period and suppressing the uncertainty, the deniers deliberately miss the point: temperature increases of 4–5°C and above are likely to be catastrophic; if we act strongly and effectively in the next decade we can radically reduce the probability of those temperature increases at modest cost.

More recently, others have tried to argue that the warming has stopped because 1998 (an El Niño year, with warmer surface temperature of oceans) was a little warmer on average than 2007 (a La Niña year, with cooler surface temperature of oceans). This confuses cycles with trends, peaks with troughs and sea temperatures with land temperatures. Further, it ignores that the last decade was the hottest since records began and that the trend is clearly upwards. But this is the kind of nonsense that some would try to peddle. There are many more half-baked attempts to try to naysay the science, but they always unravel on careful inspection. And the same has been true of more sophisticated attempts, such as those involving changing structures of humidity in the atmosphere.

The basic scientific conclusions on climate change are very robust and for very good reason. The greenhouse effect is simple and sound science: greenhouse gases trap heat, and humans are emitting ever more greenhouse gases. There will be oscillations, there will be uncertainties. But the logic of the greenhouse effect is rock solid and the long-term trends associated with the effects of human emissions are clear in the data. The arguments from those who would deny the science look more and more like those who denied the association between HIV and Aids or smoking and cancer. Science and policy-making thrive on challenge and questioning; they are vital to the health of enquiry and democracy. But at some point it makes sense to move on to the challenge of policymaking and accept that the evidence is overwhelming. We are way past that point.

Another argument accepts the science but claims that adapting to climate change is preferable to reducing emissions – this flies in the face of the scale of the risk. Examples of past adaptation that are often advanced, such as Romans making wine in the north of England, refer to temperature variations of 1°C or so and are irrelevant to the changes of 4–5°C and higher that we risk. And the speed at which it would occur would not only make adaptation extremely difficult, it would also very probably lead to massive conflict. That cannot by any stretch of the imagination be described as adapting at a moderate or acceptable cost. To invoke this argument is, quite bluntly, to be ignorant and reckless.

Playing the waiting game, hoping that 'something will come up' to enable us to stop or reverse warming, is also dangerously casual. Such explorations and research should of course be pursued; it is possible, for example, that methods of extracting CO_2 from the atmosphere on a huge scale may emerge. But do we want to bet the future of the planet on it? Would there be severe risks associated with re-engineering the atmosphere or oceans? And how and by whom would decisions be taken on these new technologies if these kinds of risks are involved?

Some people do not see ensuring the welfare of future generations as a high priority, or argue that it is best served not by tackling climate change today but by putting resources into other investments and spending the returns on the environment later. This argument is often made by economists who, seeing that people will generally choose, say, a kilogram of cheese today over a kilogram of cheese next week, conclude that they must therefore care more about their contemporaries than they do about future generations. The question of how to balance the well-being of both the current and future population has engendered an intense, sometimes quite subtle, but often confused debate between economists about the best approach to 'discounting' of the future. It is remarkable how many economic and other commentators are ignorant of some of the basic concepts, ideas and theories of discounting in economies that do not conform to the simplistic economic model, with its perfect markets and absence of uncertainty (for more of this, see Chapter 5).

Many also make the mistake of thinking of policy as being about simple investment projects in the context of given growth in the past being extrapolated simple-mindedly to the future. The scale of the effects at issue is much larger than simply whether a particular investment is worthwhile. This is a choice between strategies, one of which will, with some significant costs over the next few decades, take us to an era of continuing growth and poverty reduction with a cleaner, safer, more biodiverse and attractive world. The other accepts business as usual and thereby probably undermines and stops growth, and leads to immense dislocation and loss of life. The idea that the costs of reducing emissions (because of a small scale of risks, adaptation or discounting) outweigh the benefits is simply wrong, not only about the actual expense and rewards, but also because its approach is far too narrow in relation to the risks and misguided on the techniques of economic analysis such problems of risk require, particularly in the context of distorted or absent markets.

On a broader level, we could recognise that we have to make major decisions on whether to combat climate change by deep cuts in emissions, and that we have clear scientific advice but, rightly or wrongly, we have a sneaking suspicion that the advice may be misguided. What shall we do? There are two kinds of mistake we might make: act on the scientific advice only to find that it turns out to be wrong, or act assuming the science is wrong, only to discover it turns out to be right.

Which mistake is more dangerous? If the science is wrong and the risk of large temperature increases turns out to be low but we have made cuts in emissions, then we benefit because we will have a world that is more energy-efficient, with new and cleaner technologies, and is more biodiverse as a result of protecting the forests. There would be a loss in terms of the resources spent to achieve this which, had we known that the risks of climate change were overstated, might have been better used elsewhere. These might be net costs, but they would be very far from being disastrous.

If, on the other hand, the science is right but we have refused to act, then where are we when we realise we have made a mistake? We will have continued to emit for thirty or forty years and stocks will have reached levels where the risks of dangerous climate change are extremely high. It will be vastly expensive – or quite conceivably impossible – to back ourselves out of the situation.

It would be grossly imprudent to act on the assumption that the science is wrong *even if* the probability of it being right were fairly low. And if we add to that common-sense argument the reasonable assumption that there is a high probability that the science is right, then the argument for strong action is overwhelming.

To counter this very basic argument on risk one would have to argue that the risks from inaction are so small that risk analysis is irrelevant. That would be a far stronger and more specific position to take than one which says that there are many uncertainties remaining and the scientific case is still unclear – if it is unclear, then the risk argument is reinforced, not weakened. The argument for inaction, or for weak or delayed action, would make sense on the basis of reservations about the science only if one could assert that we know for certain that the risks are small. In the face of the evidence we now have, that is a complacent, ignorant and dangerous position to take. It is not healthy scepticism or an openness of mind; it is a denial of evidence and reason.

The arguments offered by those who would deny the case for strong and timely action are a tissue of confusions about both the science and the economics. Also, most of them back away from, suppress or trivialise the basic ethical issues. However, the noise made by deniers continues to be loud, in relation both to their modest numbers[28] as well as the poverty of their thinking. Why do they make their case at such volume, and why do they have an audience, given that their case is so weak?

The answers are largely political. Some on the right, such as those attached to free-market think tanks like the American Enterprise Institute, see environmental causes as a Trojan Horse used by those who would like to regulate and control the economy. Others on the left, in some developing countries such as India, for example, see the issue as an elitist hobby horse of the middle and upper classes who are diverting attention from more pressing issues such as poverty and redistribution.

In sharp contrast to some of the claims from the right, most of the policies we will look at are about fixing market failure – that is, enabling markets to work better. And as for the minority on the left, far from being elitist, climate change and the battle against poverty are inextricably linked. Poor people are hit earliest and hardest by climate change, and global responses can and must be designed in ways that are equitable and promote development, both because it is just that they be designed in this way, and because they will fail if they are not.

There are vested interests, particularly in coal and oil extraction industries, that see a move away from hydrocarbon-based energy as a threat. These industries will of course experience some dislocation, just as the introduction of roads and trains caused dislocation in the market for horses and carts. But the potential for dislocation is not an argument against change; while the adjustment costs from cutting emissions must be managed, the dislocation for society as a whole will be far higher if we continue with business as usual. Enlightened self-interest from those involved in hydrocarbons should lead to the support of technologies enabling the clean use of hydrocarbons, such as carbon capture and storage, and not to defend deniers and cranks. In the medium term, this is the only real option for maintaining demand for hydrocarbon energy sources (in the long term, of course, if their use continues they will be depleted).

There will be many who object to attempts to change their patterns of energy consumption – including (or especially) for transport – via prices, taxes or regulation. Costs of power, heating and cooling are major political issues which many politicians shy away from tackling on short-term electoral grounds. The response to those preoccupied with these short-run cost-of-living considerations is simply to show that continued high-carbon growth is impossible at reasonable cost, while low-carbon growth is achievable. Alternatives to a hydrocarbon-powered internal combustion engine are already available and many more will be created; similarly for power generation. Further, public and informal discussion can create much greater awareness and under-standing. All of these require both leadership from politicians and political action from the public.

There are economic, technological, social and political answers to all of these questions, but there is no doubt that moving to a low-carbon growth path will involve real economic and political costs. These costs must be acknowledged and managed, not dismissed. Their existence implies that there is potentially a large audience for someone who tries to argue that that change is unnecessary, is not worth the cost, or capable of being postponed. Many 'leaders' in countries such as the UK, the USA and Australia have been tempted to pander to these audiences. Whether or not they are succumbing to political temp-tation, no doubt many or most of them believe what they are saying (some appear to think they are saving the world from costly and unnecessary action). It is very important that their arguments are seen to be wrong: they are indeed profoundly misguided. The risks are clearly enormous, and the argument must move on to how to respond.

CHAPTER 3

How emissions can be reduced, and at what cost

If we start now and plan carefully, the costs of achieving low-carbon growth will be modest relative to the risks avoided. And we shall discover many new opportunities along the way that are likely to make costs much lower than we might now anticipate. A more secure, stable world, growing strongly with a safer natural environment and with less poverty is possible, but only if the world acts together and follows sensible economic and social policies. What should we aim for, how do we get there and what will it cost?

We will, in any case, have to find a way forward without hydrocarbons in the next century or so, possibly sooner, since they will either be exhausted or extremely expensive. Much of the world's easy-to-access oil and gas has already been located and is under development. What remains is in harsh frontier environments, such as deep ocean or the Arctic, where it is more difficult and expensive to extract, or is 'tight gas' – gas trapped in small pockets – and so on. The hurdle there is not *locating* resources – often we know they are there – but inventing cost-effective and environmentally responsible ways to extract them. In the words of Jeroen Van der Veer, Shell's former CEO, 'Easy oil and easy gas – that is, fuels that are relatively cheap to produce and very easy to get to the market – will peak somewhere in the coming ten years . . . it is the end of the "easy oil" era.'[1] We must, however, find alternatives before hydrocarbons start running out, because if we use all or most of them without capturing and storing the emitted carbon dioxide, the resulting concentrations of greenhouse gases would put the planet at grave risk. Put crudely, there are probably too few hydrocarbons to sustain growth for more than a century, and certainly more than enough to fry the planet.

What our targets should be

Unless we hold concentrations at or below 500 ppm CO_2e, the risks are high. Controlling concentrations at this level would give a probability over 95% of a temperature rise greater than 2°C, but only a 3% chance of it being above 5°C. Given that many scientists have argued that anything over 2°C is 'dangerous', even 500 ppm could be seen as too high. Actually we would eventually have to go down to 400 ppm to have around a fifty–fifty chance of limiting the temperature rise to 2°C.

The problem is that we start at around 430 ppm CO_2e, are adding 2.5 ppm every year and this annual increment is rising. Short of creating an immediate, major and prolonged world economic decline, it is impossible to make big and sustained reductions in emissions starting instantaneously, and at 2.5 ppm added each year we will be at 450 ppm by about 2015.

Looking longer term, we should recognise that holding concentrations below 450 ppm – and I focus on 500 ppm in much of this book – still carries serious risk of passing a number of tipping points, like the destruction of rainforests and the release of methane from thawing permafrost. It makes sense, therefore, to interpret 'holding emissions below 500 ppm' as eventually allowing a very long-term stabilisation below that level. The learning we have to do to hold levels below 500 ppm will tell us much about how to go further.

Just where around 500 ppm CO_2e we should aim is a matter of balancing the costs and the avoided risks. Some, profoundly misguided in my view, might argue for higher levels, say 650 ppm, on the grounds that 500 or even 550 ppm is too expensive. In the Stern Review, costs were based on an upper limit of 550 ppm CO_2e; the science now suggests that this is much too risky. We should see 500 ppm CO_2e as a more acceptable ceiling, revising as necessary, probably downwards, as we learn along the way. To go for the higher level of 550 ppm from the beginning risks closing the option of the lower level of 500 ppm, when in fact there are powerful arguments for targets below 500 ppm. While overshooting an upper limit and coming back is possible in theory, all but a minor overshoot leads to higher temperature increases which would take centuries to fall back down. Geo-engineering, such as introducing aerosols into the stratosphere to promote the reflection of sunlight, or using gigantic mirrors, or manipulating the oceans to increase absorptive capacity, or other methods to extract greenhouse gases should all be examined. But they carry their own largely unknown dangers as well as the major uncertainties as to whether they could work on scale. And who would take the decision to use them? Thus it would be very dangerous to relax our efforts to reduce emissions on the grounds that these might be a silver bullet that could get us out of our troubles and allow us to overshoot the 500 ppm threshold.

The upper limit, or target more briefly and crudely, is given in terms of concentrations of greenhouse gases for two reasons. First, because it is a single, clear number. If we talk of annual flows of emissions we have to specify a path over a period of time. If we speak of a temperature we have to indicate probabilities because we can control only emissions, and through them stocks, but not temperature – in other words, the temperature appears as an *outcome* with associated probabilities. Second, concentrations focus attention on the phenomena at hand, climate change and global warming – it is the concentration that is the basic direct cause. And they point our attention to the challenge of the flow–stock problem, with its attendant dangers of delayed action resulting in continuing flows and higher stocks, together with the great difficulties of reducing stocks once they are there. A delay of thirty years before taking action would probably take concentration levels to 525–50 ppm CO_2e, thus making stabilisation at or below 600 ppm CO_2e very difficult. That is extremely dangerous territory. The cost of delay is immense.

There is no unique emissions path to achieve a given target level of concentrations. For every particular path that achieves a certain level, there are always others that could also reach it, by doing a little more earlier and a little less later, or vice versa.[2] We should think of a given target level as being represented by a *corridor* of paths of emissions. Another way of looking at the issue is to see it in terms of the amount of total emissions that can be made in an overall period of time consistent with holding concentrations below a given level.[3] Whichever route is chosen within the corridor, all of the three paths for target levels of 450, 500 or 550 would have to peak soon: probably around fifteen years from now for 550, ten years for 500 and almost immediately for 450. In terms of absolute flows, in 2050 emissions would need to be around 30–35 Gt CO_2e for a path consistent with a 550 target, around 20 Gt for a 500 target, and 10–15 Gt for a 450 target.[4] Given that global emissions have been rising strongly since 2000 and are now around or above 50 Gt, reaching a flow of around 20 Gt for the 500 ppm target path involves cutting current levels by more than half relative to now and around a half relative to 1990 or 2000 (which, for world totals, were fairly close).

How to achieve the targets

Looking at the position in 2050 will help to see what is involved in achieving this 50% cut. We can then work back to think about the transition from now to then. In planning for these cuts we must think about the growth of output and population, as well as the sources and types of output and patterns of consumption and ways of living.

If world output were to grow by a little over 2% per annum until 2050, then it would expand by a factor of 2.5 – in other words, it would be two and a half times as big. Halving emissions by 2050 would therefore mean reducing emissions *per unit of output* by a factor of 5 – an 80% cut. If output were to grow by 3% per annum, then the growth factor would be about 3.3, and halving emissions would involve dividing emissions per unit of output by 6.6 – cutting by 85%. Of course, countries growing more quickly would have to cut emissions per unit of output by more to achieve a given absolute cut in their emissions.

Because in some parts of the economy, such as agriculture, drastic cuts of emissions per unit of output will be difficult (although substantial reductions are possible even here), those sectors where they are possible would have to see their emissions per unit of output reduced to close to zero. If the overall requirement is for a cut of 80% per unit of output and one major sector cuts by less than 80%, then another will have to have cut by more than that to reach the overall goal.

Different countries have varying levels of emissions by sector, depending on local characteristics, including the structure of production and natural endowments, but for the world as a whole, the breakdown of sources of energy emissions is illustrated on page 42.[5]

The categories used in this chart reflect the IPCC Common Reporting Framework used by the UNFCCC. 'Electricity and heat' accounts for all power and heat plants, including combined heat and power (CHP). 'Other fuel combustion' comprises mostly commercial and residential buildings emissions.[6]

GHG Emissions by Sector in 2000 (including land use change)

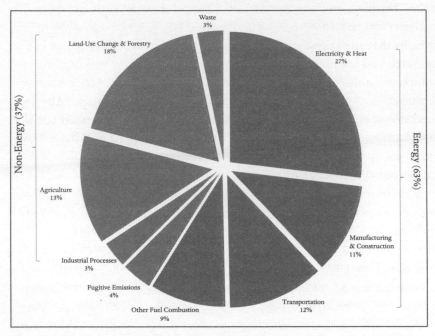

Source: Climate Analysis Indicators Tool (CAIT) Version 5.0, Washington DC: World Resources Institute, 2008.

Energy-related sectors account for around 63% of the total. Within the 37% from non-energy sectors, the 'land-use change and forestry' source is mainly deforestation and peat fires. These emissions are on a scale larger than those from road, sea and air transport put together, and more than half comes from two countries, Brazil and Indonesia. Stopping deforestation can be a major contribution to reducing emissions, and in principle it could be done quickly and at reasonable cost, as we see in the second half of the book.

Among emissions associated with energy, reductions through improved efficiency will be of substantial importance right across the board. There is great scope in most sectors in most countries. And again progress can be made quickly.

Within energy, electricity and transport contribute the most – around 60% – and together they constitute about 40% of the world total of greenhouse gases (the percentages for CO_2 only are substantially higher). There is a whole array of existing technologies which can generate electricity with very low carbon emissions including hydroelectric, wind, solar, biomass, nuclear, ground source, geothermal, wave, tidal and carbon capture and storage (CCS) for generation using hydrocarbons. It is likely that all of these will play an important role and it will be markets, research and learning, and public policy on prices, taxes and regulation, which will influence how they develop. It would be unwise to jettison or obstruct any of the possible technological opportunities. All the 'technologies' have their own problems in terms of associated social, environmental or economic costs. But problems can be managed by good policy and they are generally small relative to the magnitude of the risks from climate change.

Relative costs of producing electricity change quickly as the prices of hydrocarbons rise. With oil at $75 per barrel, and the equivalent for coal and gas, and without a carbon price, coal, combined cycle gas turbine and nuclear[7] are the cheapest options. With a carbon price of €40 ($50)[8] per tonne of CO_2, onshore wind and coal with CCS are roughly competitive with coal without CCS.[9] At $150 per barrel of oil, wind is competitive with coal and gas without a carbon price. The underlying trend of hydrocarbon prices is likely to be upwards. No doubt, there will be major fluctuations along the way, indeed in 2008 oil prices ranged from around 35 dollars per barrel to around 145. But the estimate provided in November 2008 by the International Energy Agency is for 100 dollars for the next decade rising to 120 dollars (all in real terms) over the following decade. It clearly makes no sense to make long-term investment decisions using an oil price from the middle of a recession.

Countries are likely to follow different approaches: Iceland's power is largely geothermal; Brazil and some Scandinavian countries use hydro very extensively; France's electricity is 75% nuclear; over 30% of electricity in Schleswig-Holstein in Germany now comes from wind (in some districts it is as much as 50%). While Germany is still the world leader in terms of the installed capacity in wind energy, China and India have seen average growth rates well beyond 30% over the last five years. In the USA, 35% of installed new power capacity in 2007 was wind.

However, China, India, Poland and others are likely to rely strongly on coal, perhaps 70–80% or more, for the next few decades, largely because it is cheap, power stations can be built rapidly, and it is available internally and is thus secure. (Generally speaking, the objectives of energy security and climate responsibility are well aligned. But there are technologies and resources that, being widely available, are good from an energy-security perspective, but commit the world to higher emission levels – hence the importance of CCS, not only for coal, but also for gas.)

The future structure of technologies is hard to predict from here. For example, it is possible that CCS for coal will be an interim technology for only a few decades as technological progress moves strongly in renewables and more capacity is established. Gas discoveries seem to be moving forwards and perhaps gas with CCS will complement CCS for coal as an interim technology. With prices for greenhouse gases and other policies set appropriately, this will be in large part a matter for the markets. The details of policy matter greatly and we return to these issues in Chapter 6. The technological choices will be shaped, however, by many factors above and beyond prices, including natural resources and political and social realities, including concerns about energy security. If different countries pursue different technologies, the more everyone will learn – there are gains to international diversity.

All the technologies we have mentioned are in use either on major commercial scale or as substantial pilots. But new technologies will also emerge. For example enhanced photosynthesis might transform our ability to turn the sun's energy into usable plant material. Since plants absorb CO_2 as they grow, energy based on using such material can be virtually carbon-free. The sun's energy coming into the earth is enormous, far more than we would ever need. Transforming it through solar thermal, solar voltaic or greatly enhanced plant material could give great leaps in renewable energy. We do not yet know whether electricity from nuclear fusion could become commercially viable and safe. Biomass with CCS would provide an energy source which would actually reduce carbon dioxide since the amount extracted in the growing process would not return to the atmosphere when used for fuel. If it can work on scale it could make a major contribution to reducing concentrations.

The key point here is that there are many sources of low-carbon energy, some of which we know about now – and some of these are advancing quickly. There are others of great potential which are real possibilities in the next decade or two, and further possibilities for the longer term. The speed of change can be rapid. Take the example of moving towards close-to-zero carbon electricity. France's nuclear power went from a tiny fraction to around 75% over only twenty years; Germany has expanded the share of renewable energy to 14% of total electricity consumption in the last few years, and aims to increase this to 30% by 2020.[10] With strong and determined policy and sensible milestones along the way, we can get close to carbon-free electricity at reasonable cost by 2050. And this kind of planning horizon is essential – the later we leave it, the more high-carbon sources of electricity will be locked in and the greater the cost of trying later to do too much too quickly.

Road transport is likely to expand quickly as people in developing countries acquire motor vehicles. Brazil showed how much could be achieved in two decades in road transport by the rapid establishment of biofuels after the oil crises of the 1970s – changing car engines and providing the infrastructure for the sale of different fuels. Close-to-zero carbon electricity can give us close-to-zero carbon road transport. Electric and hybrid petrol-electric cars are already in operation and are fast being improved. Technical progress in the storage of electricity through different forms of battery is likely to be swift, including radical new types like nanobatteries.[11] Storage through hydrogen has real potential too, and the infrastructure is already being built in some countries such as Sweden.

The design of towns and the role of public transport will have a powerful effect. Transport is often very wasteful because of, for example, single-occupancy vehicles and congestion, but sensible policy, such as congestion charging, can make much better use of existing infrastructure, and good physical and technological design can make infrastructure systems work much more efficiently.

Air transport will also grow rapidly – domestic flying in China, for instance, is increasing at about 15% a year and 186 new airports are to be built by 2010.[12] While emissions from air travel are currently only 3–4% of the total, this could rise – according to the IPCC – to up to 15% of global greenhouse gases by 2050 due to the increase in demand for flying. Moreover, their effects beyond standard greenhouse gases (from vapour trails) and the limited options, at least at the moment, for alternative fuels imply that air transport will be particularly problematic in the future. In the short and medium term, more efficient engines, lighter aeroplanes and higher occupancy can make important contributions, as can incentives towards these moves through pricing, taxation and regulation. In the longer term, alternative fuels with a strong power-to-weight ratio will be essential. This is a crucial research challenge, and at the moment biofuels seem the most promising contender. Boeing estimates that biofuels could reduce flight-related greenhouse gas emissions by 60–80%. An important possibility appears to be blending biofuels with existing jet fuel: Virgin Atlantic successfully tested a biofuel blend made from 20% babassu nuts and coconut and 80% conventional fuels fed to a single engine on a 747 flight from London to Amsterdam. Boeing, Air New Zealand, Continental Airlines, Shell Aviation and many other companies are quickly expanding their research capabilities in this field.

There is no doubt that 'first generation' biofuels such as corn or sugar have serious problems through increasing demand for good, well-watered land that could be used for food, but the potential for 'second generation' biofuels looks strong. They can be produced from waste products, or from crops grown on marginal land. Waste products might include straw from grains which can be converted into cellulosic ethanol or biogas from animal dung or waste, and crops such as jatropha (similar to castor oil) can bring unused or unprofitable land into profitable cultivation, provide work in poor areas and hold back desertification. We can and must learn quickly which of the various possible second-generation sources for biofuels will be most profitable, efficient, viable on a large scale and cause the fewest associated problems. Enhanced photosynthesis could play a crucial role here.

Even this very brief outline of options surely makes it clear that low-carbon growth can be a reality if we so wish. We have to do four things. The first is to make much more efficient use of energy, which is used very wastefully across the board – in buildings, industry, transport, power generation, agriculture and so on. The second is to halt deforestation. Progress on these two can be very rapid. The third is to put existing (or close to existing) technologies to work quickly. In electricity these include wind, solar, hydro, wave and tidal, geothermal and nuclear; and since hydrocarbons will be used for some time we must move quickly on CCS for coal and gas. Emissions from cars can be reduced rapidly through the design of engines and control systems, by the better use of vehicles and improved infrastructure, and by driving habits, choice of car and choice of transport; much greater use of electric cars can be made fairly rapidly. The fourth is to invest strongly in new technologies which are on the medium-term horizon, although in many cases not so far off. These include still further improvements in solar power, better batteries, enhanced photo-synthesis, new generations of biofuels, nuclear fusion and so on. The possibilities are both exciting and enormous.

Of course, the point of the journey to a low-carbon economy is that we avoid much of the danger and destruction from global warming and climate change, but there are many other desirable results from going down this path. Low-carbon sources of energy are generally much cleaner in terms of particulates and local air pollution, and are less polluting of water supplies; clean cooking fuels greatly reduce domestic pollution, which is a major cause of disease and death in developing countries; stopping deforestation protects biodiversity, while the loss of tree cover damages water retention and can have a dramatic effect on water flows, flooding and soil erosion (the floods in Bihar in the summer of 2008 were largely caused by silting of rivers in Nepal, leading to overflow, as a result of soil erosion from the loss of trees).

Energy supplies are likely to be more secure than in a world dependent on hydrocarbons, which are mostly sourced in potentially unstable parts of the world. Renewables like wind, solar, ground source and geothermal do not depend on the import of fuels, nor does much of hydroelectric power (although cross-border water issues are likely to be of increasing importance). Nuclear energy depends on supplies of uranium or plutonium which can be sourced in a number of regions many of which seem politically stable. If we find satisfactory second- and third-generation biofuels, particularly with the potential of enhanced photosynthesis, supplies are likely to be spread across the globe. The range itself gives much greater security. There would also be greater security for households and firms who could control their own supplies of solar and wind power and biogas – in many parts of the world, families and businesses are currently at the mercy of corrupt and unreliable local grids and suppliers.

The changes in technologies required to get to a low-carbon world are likely to usher in a new burst of innovation, creativity and invest-ment. Joseph Schumpeter, the great economist and economic historian, told the story in 1942 of a great surge of 'creative destruction' which can power economic growth through a whole range of new investment opportunities.[13] According to New Energy Finance, the investment in clean power in 2007 was approximately $100 billion, which accounts for almost 15% of investment in energy infrastructure. The International

Energy Agency estimates that energy infrastructure investment will average around $1 trillion per year over the next twenty years. If just one-third of that were to be low-carbon, it would be around $300 billion per year: that fraction is likely to be achieved soon and it will rise from there. There will be many types of job opportunities as new technologies and capital are developed and deployed.[14]

This is a very attractive world and it is not fanciful. It can be built using policies and technologies we broadly understand and can develop and implement. It is a world where we can realise our ambitions for growth, development and poverty reduction across all nations, but particularly in developing countries. And previous examples of rapid change in investment and technologies show that we can achieve, in the timescale that is necessary, the deep cuts in emissions necessary for a safer planet.

What it would cost

How much this transition would cost depends on when we start, and thus on how much time we give ourselves for exploiting life cycles of investment and developing new technologies. Given that we start at 430 ppm CO_2e, holding concentrations to 450 ppm means emissions would have to peak almost immediately and then fall by 7% every year for a few decades. In so doing it would probably be necessary to replace a great deal of capital equipment well before its planned retirement time. On the other hand, holding below 500 ppm requires a peak in about fifteen years, falling afterwards at about 3% a year, while holding below 550 ppm would involve peaking in twenty years, then cutting by 2–3%. Put simply, holding concentrations below 500 ppm is cheaper than 450, but more expensive than holding below 550. Achieving 500 ppm might cost 2% of world GDP per annum over the next half-century, while 550 ppm could cost around 1%. There is some uncertainty about these numbers, perhaps plus or minus 3%, but we know enough about the relevant technologies and investments to make reasonable estimates.[15]

There has been substantial analytical work on the costs of achieving different target concentration levels by asking what is possible in different sectors. Broadly speaking, this work involves two somewhat different approaches: the first we might call a 'bottom up' approach, which looks closely at individual options in different sectors but which does not attempt overall macroeconomic modelling; the second, or 'top down', works with much less microeconomic detail and tries to build an overall macroeconomic model of growth, embodying emissions. A disadvantage of top-down models is that a loss of detail on individual technologies is unavoidable: they involve modelling key elements of saving, growth, trade, supply and demand and technological change in a way that fits into a full description of an equilibrium in a world economy which is growing and changing over time – they become unmanageable and incomprehensible if too much is loaded into them. Each approach has its own contribution to make, and each of them has to make assumptions about likely patterns of future technical progress. In the Stern Review it was shown that the results from the two approaches are fairly similar.

The 'bottom up' approach is clearly and helpfully illustrated in the abatement-cost analysis produced by McKinsey, a major international consultancy firm specialising in management issues. In my view, it is this more detailed practical method using direct examples that offers more soundly based estimates than the 'top-down' approach. However, it does have its problems, as discussed below, and must be used with care. Here is a recent and detailed example of their work:

Global GHG Abatement Cost Curve – 2030

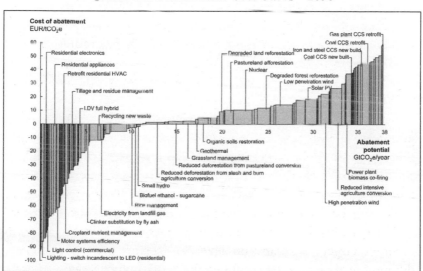

Source: McKinsey and Company. Many thanks to Jens Dinkel and his team at McKinsey for sharing this research.

On the vertical axis, the curve shows the additional cost as you increase the amount of reduction or abatement in emissions, measured along the horizontal axis, using different techniques. These techniques are ranked along the horizontal axis starting from those techniques with lowest costs and moving on to those with higher costs. The illustration in the diagram is for 2030 and has, on the vertical axis, the cost per extra tonne of CO_2e reduced, and on the horizontal axis the amount reduced in gigatonnes. The McKinsey calculation assumes a modest amount of learning about how to reduce costs along the way.[16]

Relative to business as usual, paths which hold concentrations below 550 ppm require reductions of around 20 Gt per annum by 2030, 500 ppm a little above 30 Gt, and 450 ppm just above 40 Gt; of course these broad numbers depend on what is assumed about business as usual. If we want to calculate how much an extra tonne of reduction *beyond* 20 Gt of cuts would cost, we look at the diagram at the point '20' along the horizontal axis and see that it would be around €10 ($13). Moving to the 30 Gt or so necessary for a 500 ppm stabilisation path, the additional (or marginal) cost of an extra tonne reduction would be a little over €20 ($26).

To calculate the overall cost of, say, 20 Gt of cuts, we simply add up all the marginals (moving from 0 to 1, 1 to 2, and so on) and find that the total area below the curve is the overall cost, in this case negative. For 30 Gt of cuts, this is close to zero, with the negatives (corresponding to saving money, while reducing emissions through energy efficiency) roughly balancing the positives. As we move from cuts of 20 Gt to cuts of 30 Gt, the costs of each extra tonne are in the range of €10–20 ($13–26), so the extra cost of 10 Gt (or 10 billion tonnes) is €100–200 billion ($130–260 billion).

If world GDP (currently around $50 trillion) grew annually by 2% in real terms, in 2030 it would be in the region of $75 trillion. Future exchange rates are notoriously difficult to forecast, but in euros this might be around €60 trillion, so a cost of €100–200 billion (or $130–260 billion) would represent something of the order of 0.2–0.3% of GDP. This illustrates a bottom-up calculation and shows the type of assumptions that matter (the reduction target, how fast costs fall with technical progress, what low-cost options are available, whether low-cost options are selected first, the rate of growth of GDP, and so on).

Let us allow, however, for problems in policies, markets and implementation. If, for example, there were few low-cost options available and the average cost per unit over the 0–20 Gt range of reductions were €25 ($33), then the overall cost for 20 Gt of reductions would be €500 billion ($650 billion), or 0.8% of world GDP, in contrast to the negative or close-to-zero figure using the McKinsey analysis with its big energy-efficiency options. If the extra cost per unit for the range 20–30 Gt of reductions were on average around €50 ($65), then there would be an additional €10 ($13) x 50 cost, making a total of €1,000 billion ($1,300 billion), representing around 1.6% of world GDP in 2030 for the 500 ppm strategy. Given that policy and markets are likely to be far from perfect, we might round these figures of 0.8% and 1.6% of world GDP for holding to 550 or 500 ppm CO_2e up to 1% and 2%. The overall rounding up for bad policy and nasty surprises looks fairly generous given that there will be 'pleasant surprises' too in terms of technical discoveries.

The combination of the McKinsey analysis and of emissions paths are a do-it-yourself kit for calculating costs under varying assumptions. For example, in 2050, the required reductions relative to business as usual for holding below 500 ppm are around 65 Gt CO_2e per annum. If the average price per tonne were $30, then the cost would be around $2 trillion, or 2% of world GDP (if world GDP doubled to $100 trillion between now and 2050). Thus for two different years along the path a cost estimate of 2% of world GDP looks to be in the right ballpark.

Very importantly for policy, this type of analysis can help give us an understanding of where carbon prices (or prices of greenhouse gases generally) should be. By 2030, cuts would have to be about 30 Gt for holding concentrations below 500 ppm, which suggests – if we choose to interpret the McKinsey diagram as a marginal abatement cost (MAC) curve[17] – a CO_2 price of €20–25 ($26–33) per tonne. If the price were at this level and the McKinsey diagram could indeed be so interpreted – then all options for reducing emissions which cost *below* this would be economically viable and that would cover 30 Gt of reductions. If the curves stay static, then for 2050 the price would be substantially higher. For guiding investment decisions now, a price for 2050 would not be crucial.

We must be careful about assuming these 'curves' are fixed over time. Their shape and level at any point in time will depend on past experience. The more rushed our attempts, the more we may find the whole curve shifts up. The more we delay, the more infrastructure will embody higher levels of emissions and the more costly it will be to change. Further, by the time we get to 2050, there may have been – indeed there will be, if we invest in it – substantial technical progress which will shift these curves down.

Having put the curves to use we can see both their advantages and disadvantages. On the positive side, the ordered structure of options illustrates both the importance of policy in promoting the seeking out of low-cost options, and that there *are* many options. The analysis draws attention to the importance of the negative cost options associated with energy efficiency. It also draws attention to the importance of using the price mechanism for greenhouse gases to ensure that options for reductions with cost below that price are viable and those with cost above it are not. And it allows the estimation of appropriate prices associated with given target cuts. All these insights and conclusions are of great importance for policy.

On the other hand, if crudely and naively interpreted, this analysis might lead us astray. It might be taken to imply that we have a world planner who can identify with full information and wisdom where emissions cuts should take place and make sure they happen according to plan. Or it might be taken to imply a perfection in policy and competition in a market economy which could look implausible in a world of imperfect information and special pleading. Some costs will depend on whether action takes place simultaneously or separately; for example, if there is coordinated action on deforestation, there will be less 'leakage' from one place to another and the costs of avoiding deforestation will be lower. Other costs will depend on the role of a technology within a system such as an electricity grid which must cope with peak loads as well as base-load demands. But these pitfalls arise only if we apply the analysis in a simple-minded and unthinking way. Overall, the approach provides crucial guidance, both on policy and on costs, and we have taken care to round up the illustrative calculations very substantially to get to our figures of costs of around 2% of world GDP per annum for holding concentrations below 500 ppm CO_2e, and 1% for 550.

It is possible, however, that net costs may be lower than those indicated for three reasons. First, technical progress is likely to be very rapid. Second, the associated benefits of a low-carbon economy (which we looked at earlier in this chapter) should be set against the costs. Third, most calculations use conservative long-term assumptions on the oil price, often $60 per barrel or lower for the next twenty years.[18] Relative to higher oil or gas costs, other options become cheaper, and the greater the gain from energy saving.

Consider wind energy, for example. Ever since antiquity people have sought to use the power of wind to lighten the burden upon themselves. Around AD 600, the Persians designed simple windmills to grind grain. In the twelfth century the famous Dutch windmills were used to drain water from fields. The windmills of pre-industrial Europe were estimated to provide 1,500 megawatts of power (a level not reached again until the late 1980s). In 2007, the top five countries in terms of installed capacity are Germany (22.3 GW), the US (16.8 GW), Spain (15.1 GW), India (7.8 GW) and China (5.9 GW). In terms of economic value, the global wind market in 2007 was worth about €25 billion ($37 billion) in new generating equipment. The Spanish government has a target to install a total of 20 GW of wind power by 2010 – wind farms already account for 10% of electricity generated in Spain in 2007. The US Department of Energy has set a target of generating 20% of its energy from wind power by 2030, totalling 300 GW of electricity generation. This means enhancing capacity from 2,000 installations per year in 2006, to almost 7,000 per year in 2017. The department asserts that integrating 20% wind energy into the grid can be done reliably for less than 0.5 cents per kWh. Wind accounted for 35% of total installed power capacity in the US in 2007.

The Indian company Suzlon has emerged in the last decade as a global market leader in wind power. It currently has a 10.5% global market share and has consistently maintained over 50% market share in India. It has so far installed 3,000 MW of wind turbine capacity on the subcontinent. The Indian Ministry of Renewable Energy has set a target of installing an additional 10,500 MW by 2012. In the UK, the Renewables Obligation requires all suppliers to source 10% of their supply from renewable technologies by 2010.[19] Wind energy is central to the fulfilment of this obligation and is set to account for 8% of electricity generation in the UK by 2010. The UK is currently falling behind its target, with wind accounting for around 3% of energy generation rather than the anticipated 4.7%.[20] The global wind energy market is expected to grow by over 155% to reach 240 GW of installed capacity by 2012.[21] This would mean that wind energy will account for around 3% of global electricity production in 2012 and represents an investment of over $277 billion in the next five years. The story of wind is just one example of an alternative technology growing very quickly. There will be many more.

We should be careful about thinking of electricity generation technologies one by one. Electricity generation is made up of a system of sources: some supply base power, like nuclear, some may be intermittent like wind, and others are flexible for dealing with peak supplies like gas. The combination will matter if demands that vary by season and time of day are to be satisfied without blackouts. Storage of electricity can overcome some of these problems, as can linking of sources from many different places and technologies.

Many people, including myself, think 2% of GDP per annum is well worth paying to reduce the chances of temperature increases above 5°C from around 50% to about 3%. This is a dramatic reduction of the risk of genuinely disastrous outcomes for the planet, and the cost should not be viewed as large relative to the reduced threat it buys. Such a payment is not very different from the premium to insure against a small probability of a disastrous outcome. There are many examples of governments deciding to 'purchase insurance', by, for example, putting in flood defences, civil defences, preparing against avian influenza and so on. The resources required in the case of climate change are equivalent to a one-off 2% increase in a cost or price index: it is 'one-off' because costs will be incurred mainly from using one method of generating energy rather than another; once we have made the shift, we will stay with the new methods or base technical progress strategies on moving beyond this new blend of activities. Our economies already cope with exchange-rate movements over extended periods, or changes in trade terms, which are larger than this and they still carry on growing. Another way of looking at it is to recognise that such a one-off cost would imply that the world economy would take roughly an additional six months in reaching the level of world income it would otherwise reach by 2050 (assuming high-carbon growth were sustainable for that long).

At the same time we must also recognise that the costs involved are not small: 2% of current global GDP is around $1 trillion per annum. So while these sums are justified, it is vital to keep them down. Price-based mechanisms, together with a broad coverage of sectors, technologies and countries, can cut costs sharply because markets will seek out lower costs and will find many options available. While imperfections in policy and markets in making cost estimates have been allowed for, the costs will increase if policy is bad, or if we run into more severe obstacles than envisaged. A particular worry would be weak international collaboration, which would mean that across countries or regions cheaper options for reductions would be missed, thereby increasing overall costs. On the other hand, they could decrease if technological progress is faster than the fairly modest assumptions built into the calculations. I would particularly emphasise the importance of good policy in keeping costs down, the design and implementation of which requires careful scrutiny and analysis. The analysis of such policy is central to the 'blueprint' of this book and forms the bulk of its second half.

CHAPTER 4

Adapting to climate change

The climate is already changing. To act as if the future will be like the past is simply foolish. In many parts of the world the effects of a temperature increase of 0.8°C are already proving traumatic. Low-lying island states such as Tuvalu are submerging, while about a third of Africans live in drought-prone areas, and it is estimated that by 2050, anywhere between 350–600 million people in Africa will live under conditions of severe water stress.[1] Western South Africa, Australia and parts of Mediterranean Europe are experiencing ever greater pressure on water resources, and in some cases already rely on rationing and desalination.

Even if we are sensible, past emissions will combine with those we emit in the near future and we will have to cope with probable increased average temperatures of about 2–3°C, possibly more, relative to 1850. These effects will become much more intense and adaptation will be essential and costly. We must plan ahead. While for some people these increases could create opportunities (in Northern Europe, Canada and eastern Russia, for instance, agricultural prospects could change for the better, and new shipping routes would open because of receding ice sheets), the net costs and damages will be very large. London has had to make far more extensive use of the Thames Barrier than had been predicted to combat more intensive storm surges than anticipated. The south-west of the USA is drying and the Gulf of Mexico is becoming more hurricane-prone.

There would be very few winners from temperature increases above 4°C, and major catastrophes on a world scale would be unavoid-able. Large-scale permanent migration – the most extreme form of adapting – would bring threats of global conflict, much of it over water resources. And it is happening now: conflict in Darfur is associated with extended droughts as pastoralists move in search of food for animals. Many refugees around the world should be seen as 'climate-change refugees'. I first heard this term in 2004 when speaking to Anna Tibaijuka, the head of UN Habitat in Nairobi. We were both working on the Commission for Africa and were visiting Kibera, a slum area of Nairobi where the population has grown from a very small number to about a million in only fifteen or so years.[2]

It is not easy to estimate the overall scale of mitigation likely to arise from temperature increases of 3 or 4°C and above, but given what we are already seeing at comparatively small temperature increases it is likely to be immense. It would be wrong to attribute all rural–urban migration to climate change, but over the last half-century, output per acre for most grains in sub-Saharan Africa has been stagnant while population has grown by a factor of four; some part of that agricultural stagnation is likely to have been caused by a changing climate.[3] Floods in Mozambique in 2000 and droughts in northern Kenya in 2006 are extreme events, but the region has witnessed a deterioration of environmental quality and agricultural conditions over a long period. Adaptation by strengthening African agriculture cannot wholly overcome climate threats, but it can make a big difference.

All of these impacts threaten growth and development throughout the world. To the extent that adaptation defends against those impacts, it will have a beneficial effect on growth relative to no adaptation, but we need to recognise the rising cost of investment associated with adaptation – development in a more hostile climate is costlier – and that if investment costs rise, growth rates will go down. Part of the process of adapting to more frequent disasters, which shorten the life of capital equipment, is to invest less, lowering growth rates further.

These examples and arguments make it absolutely clear that the problems of development and climate change, the two greatest challenges of our century, cannot be decoupled, separated or ranked. If we fail to control climate change we will undermine development. If we try to continue with development as if climate change is not happening, that is to fail to adapt, we will endanger development. If we try to separate projects or programmes into those which are development and those which are climate change, or even try to be precise about elements which are one or the other, we risk confusion, disruption and incoherence. We must be committed to *both* issues, analyse them together carefully, and act in a way that integrates them into projects and programmes.

Adaptation 'versus' mitigation

Though adaptation must be an essential part of the management of climate change, it is reckless to argue that it should be the predominant response. Adaptation would mute only a small part of the consequences – the cost of more frequent severe storms, typhoons, cyclones and hurricanes will still be very large, whole landscapes will still become uninhabitable because of desertification and submergence, and there will still be large-scale movements of people as a result. Adaptation is essential, but it is making the best of a bad job.

Many of those who deny the science also champion adaptation as an 'alternative' to mitigation. In other words, as well as saying 'We do not think the risks are serious', they also say 'Even if they are, we can adapt and deal with the consequences effectively as and when they occur'. These statements can be made only by those who fail to grasp the basic science and therefore fail to apprehend the consequences of the likely increases in temperature. Many of these climate 'sceptics' (perhaps 'irrational optimists' might be a better term) further imply that those who argue for strong mitigation have forgotten about adaptation. This, of course, is completely false: those who advocate strong action to reduce emissions are also those who do so precisely because they see that the consequences of climate change are with us, that more extreme weather is on the way, and that adaptation is essential. If one recognises the key messages from the science it makes no sense to present the issue as adaptation 'versus' mitigation, i.e. as one or the other, yet it is remarkable that there are still those who say essentially that adaptation should be the predominant response.[4]

Those who argue strongly for mitigation must continue to make it clear that this does not imply that adaptation is a minor issue. On the contrary, climate change poses very severe challenges to growth around the world. Without adaption, economic development and poverty reduction in poor countries will be seriously undermined.[5]

For rich countries, the challenges and costs will also be huge. Take London, the city where I was born and now work. While not the worst affected by climate change, temperature increases of 2–3°C will involve major changes and investments: wetter winters will require much greater capacity in sewage systems; the Underground system will require substantial investment in air conditioning; the increasing severity of surges in the Thames will require large investments in the Thames Barrier and flood defences generally. Adaptation to a changing climate has existed throughout history. The Darwinian theory of evolution is in large measure as much about the effect of climate and environment as it is about the mutation of species. We will focus on the developing world but there are many examples from the rich world. People in Japanese coastal regions have been protecting against typhoons for centuries, partly by making information available as early as possible, partly by investing in infrastructure such as a refuge port for boats, and partly by community action to hold boats together. A similar story of information, infrastructure and community action can be seen in the dykes, dams and canals of the Netherlands, around a quarter of which is below sea level, and which have a history of two millennia. In the last twenty years, Philadelphia, once termed the 'heat-death capital of the world', has developed systems of community support and shelters. Adaptation can do much to reduce the damage of climate change.

So while we must take on the great challenge of mitigation with urgency and commitment – and the more we do so, the less we will have to adapt – we must never lose sight of the importance of planning for and acting on adaptation *now*. A focus on mitigation must never become a 'conspiracy against adaptation'; we need to do both, on a very large scale.

How we can adapt

Adapting to the consequences of climate change requires anticipation, investment (in information, equipment and infrastructure) and organisation. It will in many cases involve radical changes in patterns of economic activity and ways of living. There are many actions which will be justified under a broad range of, or perhaps all, possible outcomes. These might be called 'win-win' strategies. A crucial example is water management. Water is allocated inefficiently around the world. Water tables are depleted in India because there is no charge for extracting ground water. Rice, absurdly, is grown extensively in California – a very dry area – because water is not priced properly. Water is wasted in many places through leaks which could be cheaply fixed. As water becomes ever scarcer in many parts of the world, through climate change, this type of waste becomes more and more inexcusable.

While the challenge of mitigation is global, adaptation must take place mainly at the local and regional level. Consequently, the scientific information guiding actions for adaptation must be much more refined, or of 'higher resolution', than the information necessary to drive arguments for mitigation. As the climate is an integrated system with many complex and detailed factors interacting with and influencing both global and local outcomes, computer modelling has to work with smaller cells; the number and intricacy of the interactions dramatically increase the level of computing power required.

To generate the necessary information, investment in computing facilities – and in the training and retention of climate scientists – has to go up. It is striking that the biggest computers in the world are military, devoted, inter alia, to the study of nuclear weaponry; the pay-off in terms of human security from allocating resources for supercomputers to the analysis of climate change would be immense.

The local climate can be powerfully influenced by local elevations, soils, hot spots, vegetation, sea currents, and so on. Long periods of observation are required to understand local phenomena. Forecasting climate at the local level is therefore much more technically complex than describing possible broad regional or global outcomes, as it involves, in addition to the regional information, the analysis of local structures and integrating these analyses with the global structures.

Major international collaboration will also be necessary in order to combine the very big models with local information and analysis. The resources required to support this collaboration could be modest relative to the value of the information it brings. If better information improved the quality of adaptation in developing countries by only 10–15%, the returns might be $10 billion a year in the near future – and these gains will rise rapidly as the overall challenge of adaptation looms larger and larger.[6] The numbers are likely to be still higher for developed countries, given the greater monetary value of the physical assets at risk. These potential returns are enormous relative to current worldwide expenditures on research in climate science, which is only a few hundred million dollars a year.

There are only about ten climate modelling groups with the knowledge and computing power to carry out the global analysis that we need. The one I know best, a world leader, is the Hadley Centre in the UK, which is engaged in a number of outreach programmes. In one, they link up with computers around the world in order to expand the number of 'runs' they can do, making their probability estimates much more precise.[7] They also work with forecasting groups at a local level, particularly in developing countries, to find ways of linking their global modelling expertise with local information.

The quality of information will have a critical influence on the effectiveness of modelling and forecasting efforts at the local level, and the poorer the country, the less likely it is to have very detailed data. The annual meteorological budget in France is around $400 million, but in Ethiopia – which is twice the area of France, and has a population that is not only larger but much more vulnerable to the climate – the budget is $2 million.[8] The 2005 G8 Summit in Gleneagles promised to strengthen Africa's monitoring capacity, but follow-up has fallen far short of commitment.

One area in which insufficient information has profound implications for global modelling, as well as for regional and local efforts to adapt, is the few hundred square kilometres of the Himalayas where the great rivers of Asia rise. Changes in this crucial area can have powerful effects on the hundreds of millions of people who depend on these rivers for their livelihoods and who are very vulnerable to flooding.

The resolution and assembly of information must be designed in relation to its potential use, and in particular focused on risks to which we can respond. Similarly, the design of adaptation requires careful analysis of the implications of the information available. Adaptation strategy should apply the principles of risk management, one important tenet of which is diversification, whether it be industrial activities, skills, production techniques or locations. Relying heavily on one industry, such as agriculture – or, worse, on one crop within agriculture – greatly increases vulnerability. Similarly, if climate change disrupts the numbers of certain species of fish in a body of water, for instance, the impact on local fishermen will be much more devastating if their boats, nets and skills are geared primarily to catching those particular fish.

Flexibility is another important principle in managing risk, and education plays a key role here, as does health – if people are physically weak they are more vulnerable to dislocation and to new or more intense diseases brought about by climate change. Improving human capital, especially through education, allows workers to move jobs more easily as the demand for different kinds of activities changes. Some industrial sectors are much less flexible than others, since expensive and long-lived capital equipment may be very specific to a particular purpose; the same may be true for related special skills.

The importance of diversification and flexibility indicate very clearly that economic development must not only be central to any adaptation strategy, it must also have the improvement of human capital at its core. Income and wealth provide resilience against new adversities, from climate or otherwise, and also make investment in diversification and flexibility easier. Development itself is the way to strengthen a society's ability to adapt. In summary, development without adaptation, and vice versa, is profoundly misguided. Just as good policy must understand the link between development and emissions reductions, it must also understand how development and adaptation are connected. A commitment to reducing poverty *must* imply a commitment to action on both adaptation and mitigation.

The design and quality of infrastructure and buildings should be a crucial part of any adaptation strategy. Just as earthquake-prone cities such as Tokyo and San Francisco have strong codes to ensure that structures can withstand earthquakes, if appropriate care is taken at the design stage, infrastructure can be made much more resilient to climate change. Irrigation systems clearly have to be designed appropriately if rainfall patterns and the behaviour of water systems will change. Roads, bridges, tunnels, electricity transmission and railways should be designed to cope with more storms, floods and droughts. In many places flood defence systems need to be prepared for a higher frequency and severity of floods. The same is true of sea walls and protection against sea-level rise, and buildings have to be designed to withstand much higher wind speeds. Wherever we look, adapting infrastructure and buildings will be crucial if disasters are to be avoided – or, at least, better handled. Here are some current examples of adaptation in various parts of the world:

Pacific Islands

- *Samoa*: Community grants to strengthen coastal resilience, and reconstruct roads and bridges to make them more resilient to cyclones. The thinking behind the locally targeted approach of the programme is that local people are often in the best position to identify points of vulnerability.
- *Tonga*: National programme to construct cyclone-resistant housing and to retrofit buildings to higher hazard standards.
- *Kiribati*: Climate-proofing of major public infrastructure, and promotion of more efficient water management.
- *Niue*: Strengthening of early warning system for cyclones, including satellite phone backup, solar-powered radios for isolated villages and email facilities. The government is also promoting vanilla as a more climate-resilient cash crop than taro, which typically suffers heavy damage during cyclones.[9]

Qinghai-Tibet Railway

550 km of the Qinghai-Tibet Railway crossing the Tibetan plateau rests on permafrost, about half of which is only 1–2°C below freezing, and is therefore highly vulnerable to even moderate warming. Thawing could significantly affect the stability of the railway, so design engineers have put in place a cooling system using crushed rocks. In winter, the colder denser air above the rock layer circulates downwards through the spaces, forcing warmer air out and away from the ground. In the summer, the air is warmer and lighter outside the rock layer, and the air within the rocks will cease to circulate, thus minimising the amount of heat absorbed by the permafrost. The technique could be applied to many types of infrastructure projects in permafrost zones around the world.[10]

Water harvesting practices in the Indian Himalayas

Communities in the Indian Himalayas are faced with erratic rainfall during spring and summer, which marks a short growing season (two to five months). Farmers have developed several water harvesting practices to ensure food security and additional income, including:

- *Small ponds*: Spring water is collected in small reservoirs scattered at intervals on the high uplands. Water can be drawn from these ponds when required. It is a common practice in cold deserts and temperate wet areas. Over time the ponds are sealed with silt and clay particles, thus infiltration/percolation losses are reduced and the life of the pond increases.
- *Roof-water harvesting*: In the lower areas of Himachal Pradesh during the rainy season, roof water is collected in dugout structures near the houses, which are known as *diggi* in the Kangra district and *khati* in the Hamirpur and Bilaspur districts. These structures are dug in hard rocks.
- *Harvesting of rainwater*: In the hills, rains are erratic and torrential and a relatively high percentage of rainwater goes as run-off and stream flow. It carries fertile soil and plant nutrients which makes the soil degraded and barren. This excess water is stored directly in the farm ponds and depressions, or the stream flow is diverted to safer points where it is stored. This water can then be used for irrigation from dugout structures.[11]

France's heatwave plan ('Plan Canicule')

Following the summer 2003 heatwave (the hottest three-month period recorded in France), which caused an estimated 15,000 extra deaths, the French government prepared 'Plan Canicule', which consists of four different levels of intervention.

1. Vigilance: Active every year from June to September to monitor action plans and keep the public informed.
2. Alert: Trigger national and regional public services when temperatures exceed critical levels.
3. Intervention: Medical and social intervention when a heat-wave is under way.
4. Requisition: Reinforce existing plans and apply exceptional measures when a heatwave is long-lasting, for example through use of government transport and army mobilisation.

The national plan is supported by a series of action plans that focus on particular vulnerabilities – (i) care homes for the elderly; (ii) medical emergency services; (iii) emergency alert systems; and (iv) Paris.[12]

Urban reforestation in Toronto, Canada

Climate change results in more frequent and extreme weather events such as summer heatwaves. These effects are exacerbated in urban areas where dark surfaces and infrastructure present in the city amplify the heating capacity of incoming solar radiation. Heat islands develop in cities as naturally vegetated surfaces are replaced with concrete and other artificial materials. Increasing temperatures pose significant health risks to residents in Toronto, Canada, including: heat-related illness and mortality during heatwaves; exacerbation of pulmonary disease due to concentrations of ozone and particular matter; and increased risk of certain infectious diseases, such as encephalitis, which spread more readily in hotter weather.

Urban reforestation can mitigate the effects of heat islands. Planting trees and greening roofs not only help to shade cities from incoming solar radiation, they also increase evapo-transpiration, which decreases the air temperature. Trees can reduce energy costs by 10–20%.[13]

For agriculture, a particularly important challenge is to develop climate-resilient crop varieties and techniques. As well as significant investment, progress will depend on international agricultural research systems and stations making it a priority. The group of bodies associated with the Consultative Group on International Agricultural Research (CGIAR) has played an outstanding role in crop development and the improvement of practices – for example, in their development and dissemination through the 1970s of the agricultural techniques associated with the Green Revolution – and can and should play a leading role in developing new, climate-resilient crop varieties and agricultural techniques.[14]

Cultivation techniques which use water more economically – in rice cultivation, say – are also likely to release fewer greenhouse gases like methane. Low-till agriculture may preserve the water content of soils, helping with adaptation while simultaneously releasing less carbon, because there is less disturbance of the soil. In agriculture, as in other activities that generate emissions, adaptation can be combined with mitigation.

A promising approach to lowering CO_2 in the atmosphere while producing energy is biochar bioenergy. Chars from the thermo-chemical processing of biomass are believed to increase soil fertility by improving nutrient and water retention, lowering soil acidity and density, and increasing microbial activity. In addition, energy produced from the thermochemical processing of biomass that stores carbon as biochar in the soil can be considered *carbon negative* due to biochar's high carbon content.[15]

Adapting buildings so that they cope more easily with higher temperatures is another response which bridges adaptation and mitigation – after all, many traditional buildings in low latitudes are designed to cope with high temperatures without energy-intensive cooling systems.

A substantial part of any strategy must also be to facilitate recovery from damage after the fact. The insurance sector is already re-evaluating the probabilities of extreme weather events and natural disasters, and as the likelihood of severe damage goes up it will be important to develop ways to share risk and reduce the exposure of the most vulnerable, who are often the poorest.

In the US, anecdotal evidence suggests that a number of large insurance companies have refused to renew homeowner policies not only in hurricane-battered places like Florida and Louisiana, but also in downtown districts of New York, such as Brooklyn, on the basis that they lie in high-risk hurricane or flood zones. According to one estimate, about 50,000 residents of the New York metropolitan region – and about one million homeowners in the mid-Atlantic and New England states – have had their policies cancelled since 2004. While most homeowners have been able to find coverage with other major insurers, or with smaller companies, in many cases they have had to pay higher rates and with larger deductibles.[16] In the UK in 2008, the government and insurers have agreed to ensure that flood insurance remains widely available in the long term. The agreement specifies that the government puts a long-term investment strategy in place, setting out prevention aims and assessing future policy options and funding needs. Insurers agree to make policies available where the risk is no worse than a certain annual threshold (1.3%) and to offer cover to those existing households and small businesses at significant risk, providing there are plans to reduce the probability to an acceptable level within five years.[17]

Extensive programmes of crop insurance can be developed to help cover farmers. These can be difficult to administer but could, in principle, be handled at a district level. Administration can be simplified and problems of false claims reduced if payments are triggered by measurable events occurring in that district. Lombard General Insurance, in association with Weather Risk Management Services, has launched an insurance product in India to cover the risk of farmers growing wheat: the idea is to link claims to an index of weather data, rather than actual crop losses, which reduces significantly moral hazard and the time it takes to settle claims.[18] Also in India, a weather insurance initiative was launched in 2003 by a group of companies called BASIX, which in its first two years grew from covering 230 farmers in one state to 6,703 across six states. It generated a great deal of interest in weather-related insurance in India, and other companies now offer a similar product. In the coastal Andhra Pradesh region, micro-insurance services have been provided to a group of women as part of the voluntary Disaster Preparedness Programme. Elsewhere, the Oriental Insurance Company offers affordable cover to poor communities through cross-subsidy with the wider insurance market. Oxfam pays half the premium.

A final important feature of adaptation is disaster management, both before and after the event. The way in which the logistics of warning and relief are handled can have a big influence on the eventual scale of the disaster (the tsunami of December 2004 would have caused much less loss of life if information had been transmitted earlier). Disaster response is much more effective if reliable and focused information can be collected quickly after the event so that assistance goes to where it is most needed. Comparing the Chinese reaction to the earthquake in Sichuan in May 2008 with that of the Myanmar government to Cyclone Nargis in the same month, we can see the difference that organisational logistics can make. For the latter the requirement for extra transport, equipment, food and medical services was denied for a long period with the consequence of substantial and unnecessary loss of life.

Preparation for many of these kinds of disasters is best handled at an international level where equipment and vehicles can be shared and made available quickly, and relevant experience successfully exploited. For an individual country, particularly a small, poor country, it can be very costly to hold the necessary hardware – Ethiopia, for example, faced a series of huge forest fires in early 2000, the severity of which could have been eased by helicopters, but the government simply did not have many.

Extra funding for disaster preparedness and management can show strong returns. In China, expenditure on flood controls of $3 billion were estimated to have returns of $12 billion; in India, disaster programmes in Andhra Pradesh have shown benefit/cost ratios of 13 or more; in Vietnam, planting mangroves to protect coastal populations from typhoons and storms yield cost-benefit ratios of 50 or more.

The costs of disasters and extreme events related to climate change will of course still be very large, but it makes good sense to prepare and protect as best we can. This should be a high priority for national policy and international assistance.

The greater challenge of development in a more hostile climate and the costs of adaptation in developing countries

In his special contribution to the 2007–2008 Human Development Report (HDR) published by the United Nations Development Programme, Desmond Tutu notes, in relation to adaptation, that 'Perhaps the starting point is the inadequacy of language'. He was referring to the difference between, on the one hand, more air conditioning in a rich country, and the uprooting and migration of families in a poor country on the other. Both might be called 'adaptation'. As a matter of concept and language, I prefer 'the greater challenge of development in a more hostile climate' to 'adaptation'. It is wordier, and 'adaptation' might function as shorthand, but the fact that adaptation and development are inextricably linked needs to be clear; we must not imply that they are distinct. In human terms the costs of adaptation are likely to be much larger in poor countries.

The costs of redesigning or adapting infrastructure are likely to be higher for richer countries, in per capita money terms, since their asset base is larger. Given that wealthy nations not only have more resources and better technologies but are also historically responsible for the bulk of emissions, they should play a major role in meeting the extra cost of development in a more hostile climate, regardless of where these costs are incurred geographically. It is the poor countries that will have to shape and decide their strategies.

After alluding to the problematic use of language when talking about adaptation, Archbishop Tutu drew attention not only to the moral obligations of the rich world, but also to its self-interest. He said that we risk a drift in the short or medium term into an 'adaptation apartheid' in which the rich world adapts in the short run by spending more on flood defences and air conditioners, while the poor world, unable to undertake the necessary expenditures, faces increased hardship and risk. He went on to say:

> Allowing that drift to continue would be short-sighted. Of course, rich countries can use their vast financial and technological resources to protect themselves against climate change, at least in the short term – that is one of the privileges of wealth. But as climate change destroys livelihoods, displaces people and undermines entire social and economic systems, no country – however rich or powerful – will be immune to the consequences. In the long run, the problems of the poor will arrive at the doorstep of the wealthy, as the climate crisis gives way to despair, anger and collective security threats.[19]

Analysis and evidence on the overall extra cost of development as a result of climate change is still in its early stages.[20] To date, while it is inevitably based on sparse data and simple assumptions, the best estimates of annual costs of adaptation in the near term are from this same Human Development Report, and from the IPCC Fourth Assessment Report 2007. According to the HDR, $86 billion per annum will be required by 2015 – $44 billion for making investments more climate-resistant, $40 billion for strengthening poverty reduction strategies and $2 billion for additional disaster relief.[21]

Approximate cost estimates for making investments more climate-resistant can be derived along the lines of the following example. Overall investment in developing countries is around $2,000 billion; suppose 10% of this is vulnerable to climate change, and that an additional 20 cents per dollar spent are needed to make the investment resilient: the extra investment would be around $40 billion. Different assumptions on which investment activities are vulnerable and what the extra costs will be would obviously imply different results. As investment grows, and the proportion of investments which are vulnerable rises with an increasingly adverse climate, the extra percentage costs for those investments also rises – so these numbers are likely to grow quite rapidly. Also, this illustrative calculation does not include the potentially huge costs associated with strengthening existing structures.

The $40 billion to strengthen poverty reduction is needed to compensate for the increased difficulties that poor people will face in their everyday lives as a result of climate change. In large areas water will become scarcer and crop production will be affected; many poor people, but particularly women and girls, will have to spend more time and energy fetching water from further away.[22] The problem is not just lost time in itself, important though that is; it makes a girl's attendance at school less likely. The costs of achieving any given development goal, whether it be water availability, girls' education, or human and economic development more generally, will rise. It is difficult to provide calculations of these costs, but the HDR's estimates are the best we have.

These numbers, for the most part, are likely to underestimate costs in a crucial way: much of the threat is to lives and to health. For example, if girls or women have to travel longer distances to get water, they are more likely to be attacked. The extra costs to human life and health are not explicitly included in the calculations which, to the extent that they are only implicitly included via resources for social protection, are therefore very partial. While we do not know precisely what these extra costs are, they are likely to be in the tens of billions of dollars per annum, and rising. Health expenditures in developing countries total hundreds of billions of dollars per annum, and are, in any case, inadequate at current levels. As a few hundred million die in the developing world each year, even small percentage increases in mortality rates as a result of climate change can lead to rising human costs of great magnitude. The people of the developing world will bear these costs, but the cause, in terms of climate change as a result of greenhouse gases and their past responsibilities for emissions, surely indicate that the rich world should make a major contribution.

The sums involved are not small in relation to Overseas Development Assistance (ODA). The sum of $86 billion compares with around $100 billion for ODA – though recent ODA figures are swollen by the falling value of the dollar (since most aid comes in other currencies), and debt relief to Iraq, which is accounted for, rather misleadingly, as aid. Annual foreign direct investment to low- and middle-income countries together is close to $300 billion, although more than 90% of this goes to the latter and not the former. Currently, the member countries of the Organisation for Economic Cooperation and Development (OECD) give around 0.3% of their GDP in ODA; if they devoted 0.7%, as many of them have promised to do by 2015, there would be an extra $150–200 billion a year.[23] If the costs of climate change to the developing world are in the ballpark of the HDR estimate of $86 billion per annum – which, it must be remembered, is conservative – then they would eat up most of the increase that has been pledged by 2015. And we should recognise that on the basis of current trends, many rich countries are likely to fall short of the target of 0.7% by 2015, while others, including the USA, have not even made such a promise.

Current allocations to adaptation funds in 2007 (total, not per annum) were $279 million.[24] This is minuscule in relation to the needs. The scale of the tasks involved in adapting to climate change, and the intricate links between adaptation and development, are such that a serious international contribution to taking on the problems of development in a more hostile climate must include a substantial increase in ODA. The total sums for support for adaptation in developing countries, the sources of funding including via carbon revenues and the methods of providing and delivering support for adaptation will be key elements of the global deal which will be discussed in Chapter 8. A brief discussion of some basic principles to guide that support is the subject of the rest of this chapter.

How the international community can support adaptation in developing countries

First, the rich world should take a lead on mitigation. Adaptation will be necessary on a major scale, but the stronger and the more timely the mitigation, the less will be the challenge of adaptation. Second, the rich world should provide substantial funding for adaptation. Third, there will be specific types of activity which have strong elements of knowledge, technique and logistics for which an international provision will be very important. These include, for example, information, disaster management and crop development. Fourth, the ways in which support and funding are organised from the rich countries and international institutions should ensure that the response to the challenges of adaptation and development are fully integrated.

Climate change runs across all society and the whole economy, in terms of both mitigation and adaptation. There is a central role for environment ministries in making and implementing policy on climate change, but the work cannot be left only to an environment minister. It must involve, and strongly, the two decision-making centres in most countries charged with looking across the economy: the prime minister's or president's office, and the Treasury or finance ministry. In many countries, the planning ministry also has an important cross-cutting function.

Support from outside for adaptation or special assistance, for new agricultural techniques or to avoid deforestation, for example, should be integrated into development funding more broadly. Small adaptation funds arising within a narrow agreement between environment ministers at the UNFCCC are unlikely to be of a sufficient scale. Such funds do have their value in pioneering new methods or approaches, but coordination of development efforts would become still more confusing for both donors and recipients if there were completely separate adaptation and funding channels with separate organisational roles and structures. Attempts to build strongly separate adaptation funding arrangements and institutions would be very damaging.

Buildings, transport, infrastructure and urban design must all be simultaneously climate-resistant, energy-efficient and low-carbon. New varieties and technologies in agriculture can provide substantial results for both mitigation and adaptation. International action on adaptation must also make sure that links with mitigation policies are fully integrated right across the board.

In France at the OECD, the world's development agencies, working with developing countries, set out in 2005 principles for coordination in what was named the Paris Declaration. In the jargon of development institutions, it refers to five key areas: 'ownership' – partner countries exercise effective leadership over their development policies and strategies, and coordinate development actions; 'alignment' – donors base their overall support on partner countries' national development strategies, institutions and procedures; 'harmonisation' – donors' actions are more harmonised, transparent and collectively effective; 'managing for results' – managing and implementing aid in a way that focuses on the desired results and uses information to improve decision-making; 'mutual accountability' – partner countries and donors enhance mutual accountability and transparency in the use of development resources, helping to strengthen public support for national policies and development assistance.[25] Jargon aside, the principles are sound and should be at the forefront in thinking about and managing the integration of adaptation funding and support with development.

It is clear that climate change was not taken sufficiently into account in the discussions and analyses of, for example, the funding for the development goals at the UN Conference on Finance for Development in Monterrey in 2002, in the commitment made at the G8 Summit in Gleneagles in 2005 to double ODA by 2010, or in the EU commitment that year to raise it to 0.7% of GDP by 2015. Should we simply try to add the 'extra costs of adaptation' to previous commitments? There is an obvious sense in which the rich world should do exactly that: it undertook to play a strong role in providing resources in pursuit of agreed objectives, and the amount of resources required has now risen – in large measure as a result of rich countries' actions.

At Copenhagen in December 2009 we shall be just five years away from 2015. Meeting existing ODA commitments by then will probably be a major fiscal challenge in many countries. Of course, they can easily afford more than 0.7% of GDP, if they choose to: this figure is already exceeded by Denmark, Luxembourg, Norway, the Netherlands and Sweden. But those who are far below will need time and effort to get there. Rather than insisting on some hard-to-define 'additionality' of adaptation funding it may be better to argue that the rationale for this level of aid is very powerful even before we understood and considered the effects of climate change; when we factor climate change in, and it would be foolish to ignore it, the arguments are absolutely compelling.

As we look beyond 2015, as soon we must, and reappraise our commitments, it would be grossly negligent to again leave climate change out of analyses of strategies and funding requirements. The two great challenges of the twenty-first century, fighting world poverty and tackling climate change, must be tackled as an integrated whole by a united world.

Developing countries feel a powerful and understandable sense of injustice. They were not responsible for the majority of greenhouse gases. They will be hit earliest and hardest. And now they feel they are being told to follow a path to prosperity very different from that followed by rich countries – a path for which some argue there is as yet no model. They will surely find any deal ethically unacceptable unless it embodies strong collaboration on adaptation, with the funding to support it. In discussing both the potential devastation from climate change and the importance of mitigation, Desmond Tutu concluded:

None of this has to happen. In the end the only solution to climate change is urgent mitigation. But we can – and must – work together to ensure that the climate change happening now does not throw human development into reverse gear. That is why I call on the leaders of the rich world to bring adaptation to climate change to the heart of the international poverty agenda – and to do it now, before it is too late.[26]

CHAPTER 5

Ethics, discounting and the case for action

Action on climate change requires the current generation to make decisions about the reallocation of resources that will have profound implications for future generations, for distribution within generations, including among those living now, and for the planet and its entire species. These issues have to be examined and discussed directly if we are to have principles, objectives and criteria to guide decisions. Without such guidance, whether or not it can be expressed formally or quantified, it is not possible to assess the pros and cons of different strategies.

Also vital is an explicit analysis of ethics which, although it cannot itself determine the social values to be applied, can help identify the relevant issues and recognise inconsistencies. It is remarkable how many economists make the mistake of shying away from this discussion, engaging instead in fruitless attempts to identify so-called 'revealed' ethics and values from the outcomes of real-world markets. While there may be some narrow issues for which such an approach could be useful, climate change is not one of them. It is necessary not only to base our principles for decision-making on ethical discussion, but to do it with care – all too often economists behave as if ethical discussions are irrelevant, for somebody else, obvious, or all these things. There will not be a single answer – reasonable people can and do differ on ethics, but there should be a rational discussion.

With an understanding of the ethics, of the risks from uncontrolled or badly managed climate change, and of the costs of reducing emissions and concentrations, we have the basic materials for examining the question of the appropriate *level* of action in terms of controlling emissions and concentrations. I have essentially already proposed an answer to that question using a direct and simple route, by examining the risks from different concentrations in Chapter 2 and the costs of action in Chapter 3. With this information, I asked whether the insurance (in the form of reduced risks of damages from climate change) from holding concentrations below 500 ppm CO_2e was worth the cost, and suggested that the reply should be a clear 'yes'. However, the explicit treatment of ethics allows a somewhat more formal and model-based discussion of how to balance the costs of inaction (failing to act to reduce risk) against the cost of action. Since many economists have explored this more formal route, it is important to understand their analyses and conclusions – that is the first task of this chapter. Having done so, I conclude not only that the direct, risk-reduction, insurance-based approach adopted in preceding chapters is indeed the most transparent, robust and reliable way to think about this problem, but also that the case for holding concentrations below 500 ppm CO_2e is justified.

Over the last fifteen years or so, analysis of the interactions between the scientific, economic and ethical dimensions of climate change has commonly used what are called Integrated Assessment Models (IAMs). These combine formal modelling of climate processes with an economic model of growth, and include criteria – usually expressed in terms of a 'social utility function' – which are incorporated for assessing whether changes are viewed as improvements or deteriorations. A serious difficulty that arises when adopting this approach is that, in order to simplify scientific, economic and ethical questions to the degree required to make them amenable to this type of formal modelling, key elements of the story must be set aside. For example, many IAMs have traditionally not addressed risk at all, or alternately have done so only in the most simplistic ways. The magnitude and detail of the risks have, in all too many cases, been suppressed. Yet, as we have seen, risk is at the heart of the story. It is these kinds of simplification in some of the leading models that have led some economists to suggest that we could take a fairly relaxed view of global warming – first, that we could and should wait and see before taking strong action; and second, that we should set fairly modest ambitions for emissions reductions. Many early models also viewed ethics – often buried in mistaken appeals to market information – in a way which put a tiny value on the major social consequences of climate change in the second half of this century and the beginning of the next.

Since these results are often widely quoted as 'a view from economics', it is important to understand how and why a number of high-profile analyses have got things badly wrong. Essentially, those who would oppose action by absurdly denying the science have sometimes been joined by others who argue for weak action on the basis of the work of some economists. While these economists accept the science, they are fundamentally misleading and often simply incorrect in both their application of economics and their representation of the conclusions of the science. Fortunately, as the science has become ever clearer about the risks and the scientific models more robust, the discussion in economics has also – albeit belatedly – begun to move forward from the earlier, dangerously complacent conclusions of some contributors.

The ethical values

Different strategies on climate change will have profoundly varying consequences for different generations. We can disagree on the right ethical position to take, but there is no getting away from the fact that making policy towards climate change unavoidably requires one to take a stance on ethical questions. Analytically, it is important to investigate a range of different assumptions and perspectives on ethical positions.

No single profession or group has a monopoly of wisdom or a unique understanding of justice. While economists should not claim any special position in deciding on social value judgements, they do have an important role to play because they are used to working out the implications of different assumptions about objectives for policy decisions. In other words, while they cannot settle the question of which ethical positions or value judgements are appropriate, they can inform discussions by working out and explaining the consequences of each.

Given this, it is worrying that so many economists shy away from engaging in these discussions, or even, astonishingly, allude to those who do so as adopting 'lofty' or even 'British Empire' positions.[1] Many economists feel more comfortable staying in the world of prediction or estimation on the basis of structural models or arguments, feeling that the introduction of explicit normative assumptions takes them outside their professional territory and into the space which should be occupied by politicians or philosophers, or priests, rabbis, imams or pandits. Nevertheless, the form of the questions, and the perspectives brought to respond to them, will unavoidably, and sometimes to a large degree, be shaped by those who are trying to combine the structural and ethical analyses – that is, mainly economists. They therefore have a special obligation to be open and transparent on these issues.

Ethical values are typically incorporated into economic modelling in a very narrow way. The level of social achievement at a point in time is generally measured by a 'social utility function' or 'social welfare function' that depends only on society's total consumption, population, time and a couple of 'ethical' parameters specified by the modeller – these can be varied to examine how results might vary for different parameter values. However, we should start with a much broader perspective before narrowing like this.

Rights, responsibilities and sustainability

Two very important and interrelated perspectives concern rights and responsibilities on the one hand, and sustainability on the other. The first poses questions about responsibilities to future generations or, some would argue, to the planet as a whole and all the species that live on it. As future generations are not directly represented, current generations have to take decisions on their behalf. Part of the discussion would concern the importance of current generations' standards of living and freedoms in relation to our own actions. Another part would be to try to imagine what might be the most important elements of well-being in the future, by putting ourselves in their position.

What do we think about the importance of future generations? Do we treat them as human beings with rights equal to our own? Many political philosophies, indeed political constitutions, have a strong emphasis on the equality of rights in key dimensions such as voting, freedom of expression or the law. Are there equivalent or similar rights for those who will be born tomorrow? It is difficult to think of arguments as to why future individuals, if we are convinced they really will exist (or if we do not alter their mortality) have fewer rights than we do. We may think they will be richer, and thus better able to deal with a given loss than we are, but that is a very different issue. Of course if, when evaluating gains and losses to different groups, we go down this route of comparing income and wealth, as many would, myself included, then we would also have to consider the possibility that future generations might be poorer. This possibility becomes increasingly likely under scenarios where emissions are not cut.

The notion of sustainability is related to this 'rights approach', and argues that we should 'sustain the opportunities' of the next generation so that their freedoms are no less than those we enjoy. If we do not do this we are sacrificing their welfare in favour of our own, and putting them in a worse position than that which we have inherited. This might be seen as 'unjust'; thus their rights, in this sense, are defined by a comparison with our own freedoms. In order to apply this notion of sustainability, however, we first need to decide how we compare the circumstances of future generations to our own. It would generally be impossible to ensure that there are no reductions on any dimension; we would almost certainly, for example, give them more knowledge of technology than we were given by our forebears, and less of the non-renewable natural resources that we have used up.

A linked but narrower notion of sustainability concerns 'stewardship'. This concept suggests we have a duty to preserve key aspects of our planet, such as the environment and other species, for our descendants to understand and appreciate. In other words, we do not have complete and unambiguous ownership of what we find but are, at least to some degree, both enjoying it and looking after it for future generations. Many would think that the preservation of key sites of natural beauty, for example, falls into this category – hence the impulse in many countries to create national parks. In this approach we see ourselves as 'custodians' or 'stewards'.

All of these perspectives would have some appeal to different people, some of whom might trace them to particular religious positions and precepts. There is no doubt that many people who feel strongly about environmental issues arrive at their convictions from routes very different from that of conventional economics.

The standard economist's approach to 'discounting'

Much of conventional economics sees the value of opportunities in terms of the utility they might bring to individuals as consumers. From this perspective, the value of a good or service is determined by society's 'willingness to pay' for it and those who are taking the decision whether or not to pay are the current generation. Economists often move quickly from there to simple models in which total social welfare is calculated by adding together the utilities of each member of the current generation, and then adding the weighted utilities of each member of every future generation, where the weights used are determined by the extent to which we 'discount' the welfare of future generations.

From this introduction – in particular, the fact that the costs must be shouldered by us, whereas future generations get the benefits – it is obvious that the extent to which we discount future generations' welfare relative to our own will have a huge impact on what we conclude as to the desirability of action to combat climate change. In economics, a key but different question concerns the discounting of consumption: if an extra unit of a good becomes available later, how much less (or more) is it worth relative to an equivalent extra unit of that good now? Discounting, in relation to this good, is the process of adjusting the future value of that good to the value it would have today. The ratio between the value of an extra unit in the future and its value now is the 'discount factor' for that good. The magnitude of this ratio would normally depend on the good in question: if, for example, environmental goods and services become scarcer, their value could rise. And it would depend on who is giving up or recovering a good – losses to poorer people may be assigned higher value. The rate at which the discount factor decreases is the 'discount rate': it would normally be different for different goods, going to different people, in different circumstances.

Economists often simplify by focusing on overall wealth, income or consumption (rather than on the consumption of individual goods) and by looking at the circumstances of generations as a whole (rather than at the circumstances of individuals within each generation). In such models, they see two main reasons for giving less weight to an extra unit of income or consumption in the future than to one today.

The first is because future generations may be wealthier. The idea is that there is 'diminishing marginal utility' of income – as you become richer you attach less significance to an extra dollar. This is, of course, a value judgement but many, including myself, would regard this as an acceptable reason for discounting. There remain, however, difficult analytical and conceptual issues concerning the implementation of an approach which discounts on this basis, as we are faced with the challenge of forecasting the wealth of future generations and relating this to our actions now. What we do now on climate change will transform the circumstances and income of future generations and this will determine discount rates. It is therefore a serious mistake to see decisions in this context as being determined by a discount rate. There is also a powerful causation running in the opposite direction.

Moreover, we have to think through the ethical question of just how much we should mark down or discount increments of wealth or consumption for this reason. Broadly speaking, attempts to respond to this question usually adopt the perspective of social utilitarianism in its narrow form of adding social utilities where, often, those utilities depend only on our own consumption. This kind of discounting on the basis of growing consumption was central to the work of the Stern Review.

One way to think this through is the following 'thought experiment'. Imagine that person A has five times the resources of person B. Because B is poorer than A, we might want to give a higher social weight to increases in income to B than to increases in income to A. But how much higher? A commonly used approach would be to set the social valuation of increases in income to A relative to increases in income to B equal to the inverse of their relative wealth. In other words, since A has five times the resources of B, increases in income to B are 'worth' five times more to society than increases in income to A. Such an approach has very concrete implications for the kinds of redistributional policies we would support. It implies, for example, that society would be better off if we took £100 from A, lost or burned £70, and gave the remainder to B (the £70 is chosen at random – the statement is true of any value lower than £80, as this would leave over £20 for B, or, in other words, over one-fifth of what has been taken from A). Of course, there is no suggestion that resources *should* be lost or burned – the purpose here is to indicate the strength of the antipathy to inequality which is adopted.

Suppose, however, that we think the approach just outlined is insufficiently egalitarian. Indeed, this is precisely what many commentators on the Stern Review, such as Partha Dasgupta, William Nordhaus and Martin Weitzman, have argued. Then we might want to adopt another commonly used, but more egalitarian, approach – namely, to set the social valuation of increases in income to A relative to increases in income to B equal to the inverse of the *square* of their relative wealth. In other words, since A has five times the resources of B, increases in income to B are 'worth' *twenty-five* (five squared) times more to society than increases in income to A. This approach would imply that we could take £100 from A, lose or burn any amount lower than £96, give the remainder to B, and still make society better off (since B would be receiving more than £4, or one twenty-fifth of what has been taken from A).

Many other approaches to distributional values are possible. However, these two examples illustrate how one might think about these issues in very stylised cases. The relationship between these examples and the 'rising consumption' argument for discounting can be seen by thinking of A and B not as different individuals but as different generations – if we expect future generations to be richer than ourselves, we can think of ourselves as B, and of future generations as A. There are, of course, many complications, including many types of goods, many types of wealth, disincentive effects of transfers, the relation between this type of hypothetical circumstance and social decision-making in practice, and so on. It will always be wise to consider a range of ethical approaches and, within any one of them, different versions.

The second potential reason for giving less weight to income or consumption accruing in the future is 'pure-time discounting'. This is very different from the first, 'rising consumption', argument for discounting. It says that we should discount benefits to future generations simply because they are in the future. According to this rationale, a person with a later birthdate is, for the purposes of formulating priorities for social decision-making, 'valued' less than a person who is born earlier but who is identical in all other respects (for example, in their income stream and consumption path): an increase of consumption at age twenty to the one born later has a lower social weight than a corresponding increase at age twenty to the one born earlier. In this sense, if a pure-time discount rate of 2% is selected,[2] then a life that starts in 2010 would be assigned approximately twice the social value of a life that starts in 2045. If we were to have applied these values consistently over time, a life that started in 1970 would have been, and would continue to be, assigned twice the value of one starting in 2005.[3] In other words, someone born later counts for less. In effect, this is discrimination by date of birth. Surely many would find this very difficult to justify.[4]

One of the very few plausible arguments I am aware of in favour of pure-time discounting concerns the 'distance' of the current generation from future generations. Some, like David Hume, have argued that humans, quite rightly in his view, attach more weight to the welfare of their families than to other members of society. And one can imagine arguments which suggest that societies might function better if people and groups behaved in such a way. Wilfred Beckerman and Cameron Hepburn have discussed the use of this idea to justify pure-time discounting in models of climate change impacts, since our far-distant descendants are 'further' from us than our children or compatriots. But how relevant is Hume's 'familial distance' argument for collective decisions, as humans on this planet, about the viability of the world we pass on to future generations? Both an evolutionary and a 'functional' explanation, as in Hume, of the 'familial distance' approach focus on behaviours and physical circumstances that enable survival in competition with others on the same planet. This kind of approach seems of limited relevance to a collective decision concerning the future of the whole world.

One rationale for pure-time discounting that was incorporated into the Stern Review comes from uncertainty about whether future generations will exist. If there is a danger of, say, a meteorite colliding with the planet and destroying the world, then we might 'count' increments to welfare a few decades from now a little less heavily to account for the less than 100% probability that they will be experienced. Note that the probability of human extinction in this example refers to something outside the control of the decisions and strategies under examination. If these strategies – and it is an unavoidable question in the context of climate change – are a matter of life and death for many, then the issues are different. The simple utilitarian framework behind adding the social utilities of all living members of a society ('the greatest good to the greatest number') becomes still more problematic when it is used for taking on life or death issues, in other words decisions with profound implications for who it is that will be alive. Such issues arise for population control, or health policy, or war, or risks of road accidents, and certainly for climate change. Do we want to attach more weight to securing the life of a cheerful person on the grounds that they deliver more 'social utility' than a miserable one, or to the young over the old on the grounds that they have a longer delivery period for 'social utility'?

This kind of issue cannot be settled by tight logical argument and evidence, but there is absolutely no excuse for avoiding them, as they are central to the challenge of formulating policy on climate change. Openness and clarity of discussion is vital and economists do themselves and their profession a disservice by not confronting them. We can, indeed, talk about these issues in ways that are helpful for suggesting which values might be appropriate in making social decisions. One way of doing this, which moral and political philosophers continually use, is the 'thought experiment', simple examples of which were given a couple of pages ago in discussing income transfers and pure-time discount rates. If we take simple circumstances which we can understand and then cross-question ourselves and each other, we can make progress in clarifying these issues.

We can also look for social, community or private decisions of these types in real circumstances to gain further evidence on 'implicit' social values from a less abstract environment. However, the problem with this approach is that the answers you get depend on what you assume about decision-making processes and the perceptions of those involved concerning the way in which the world actually functions. Thus, for example, if you tried to work out what people think about income redistribution by looking at income tax structures, you would have to ask how the structure is decided and what is assumed about disincentives. When you try to 'back out' or infer social preferences from social decisions, the answers are all over the place. For example, within a given country and period some tax and transfer arrangements suggest that there is very little aversion to inequality, while others suggest a much stronger aversion. Governments can appear extraordinarily cautious with some aspects of human health but not with others, expending large resources, for example, to try to reduce probabilities of mad cow disease from very low to extremely low levels, while investing, in proportion to the harm, comparatively little in combating alcohol abuse. Or governments might regulate strongly to prevent loss of pensions (perhaps rightly) while encouraging huge speculations on house prices. One could go on, but the point here is that social decision-making in practice can be a very confusing and uninformative route to discovering social values for the challenge of long-term collective decision-making on climate change.

My interactions over several decades with philosophers, policy-makers, economists and many others, together with my own examination of the problem of the empirical inference of social values from social decisions,[5] have left me strongly inclined to the thought experiment as the simplest and most transparent way of trying to get to grips with social values. It is the approach to which moral philosophers have continued to return, most notably, John Rawls in his *Theory of Justice* (1971). It is, of course, a tool that is capable of going way beyond the narrower utilitarian constructs of the last part of our philosophical discussion.

Before leaving the standard approach within economics, and its logical problems, I should note that one key argument for action to combat climate change requires only weak ethical assumptions. That concerns the inefficiency of the market failure arising from the fact that people and companies do not have to pay for the damages caused by their emissions. Its relevance here is that we could make the next generation better off without making ourselves worse off by leaving them a combination of a better environment and less of other resources that they would be very likely to prefer to the alternative of a badly damaged environment with somewhat higher, at least for a time, other resources. Markets, uncorrected by policy, distort our decisions away from environmental goods and services; if the failure is corrected carefully, all generations could be made better off. This argument is important, but, given pervasive market imperfections, limitations on information, tax instruments and so on, we should not imply that inter- and intra-generational comparisons and value judge-ments can be avoided. They cannot.

Problems with the argument that ethical values are revealed by markets

Some economists try to evade ethical questions by arguing that social values – concerning distributional issues, for instance – are 'revealed' by the outcomes of real-world markets. It is especially common for economists to use as a reference point the rate of return or rate of interest that we find on markets (whether for investment projects or consumption or other loans), and to argue that these provide us with a reading of how much people value returns in the future relative to now.[6]

This approach can provide natural yardsticks for social decisions concerning a broad range of questions. For example, if a country with limited funds is considering building a wider bridge over a river in a particular town, then the relevant question is indeed whether or not a new bridge is likely to generate returns in excess of those obtained from putting the same funds into alternative investments. Asking whether the net discounted returns from the bridge are positive, using as a discount rate a rate of return derived from that available in other projects, essentially compares the bridge project with those other projects.

In looking at climate change, economists often argue that market rates of return can be used to infer the underlying intergenerational values that this generation actually does have. These are the values, it is then asserted, that actually should be applied to intergenerational distributional questions. To extend this kind of reasoning from the bridge-building example to the analysis of challenges such as climate change is, however, mistaken. It makes a whole series of analytical mistakes: it is simply poor economics. However, because this extension is so widely made, it is important to understand why it is wrong.

Economists who make this argument often choose a market rate of return of around 6%, which corresponds roughly to the average long-run rate of return on risky financial assets such as shares. The effect of this argument and choice of discount rate is to give a very low weight to benefits to, and costs imposed on, future generations. If one discounts future costs and benefits by 6% per annum, any damages around fifty years from now will be given a value eighteen times lower than an equivalent amount of damages incurred today; for one hundred years, the figure is 339. In other words, a cost saving one hundred years from now is given a value of 0.3% of a cost saving today. Since these far-future damages (which constitute the *benefits*, or 'avoided costs', of reducing emissions) are then compared with the *costs* of reducing GHG emissions, which must be incurred in the (undiscounted) present, it is highly likely that any model adopting this approach to discounting the future will conclude that the case for action to reduce emissions is weak, since the benefits of mitigation start to come through a few decades from now. With this kind of heavy discounting the detailed assessment of damages hardly matters.[7]

Here are examples of the series of serious analytical mistakes in the economics of this approach. First, and this is crucial, we are not thinking about small changes here but are trying to choose between strategies which have profoundly different implications for growth paths. In the bridge-building scenario, it is possible to apply a simple discount rate only because we implicitly assume that the country would continue down the same broad growth path regardless of whether or not the wider bridge were built. By contrast, when we think about different strategies for combating climate change, we are thinking about different growth paths for the planet as a whole. If future generations will be much worse off than ourselves as a result of our neglect of the problem – which is a real possibility – then an extra unit of welfare to them might be *more* valuable than an equivalent unit to us. This would imply a *negative* discount rate. This kind of radical change in growth paths cannot arise in the simple bridge-building case and it is only because of this that the straightforward application of a discount rate is appropriate.

Second, let us imagine, however, that we fail to notice that we have fallen into this logical elephant trap and continue to seek a market rate of return to ground our social discount rate. We quickly encounter a further problem. The impacts of the decisions we make *now* on climate change will start occurring in two or three decades and will last for two or more centuries. There is no financial or other market of any substance which reveals our choices as a group over a century or two. Individuals rarely have borrowing or lending contracts lasting more than thirty or forty years, while the time horizons for investments by firms are rarely more than fifty years. More broadly, there are no markets which can reveal how a generation, faced with the prospect of inflicting massive changes on future generations, should behave. Markets deal mostly with returns to individuals or firms within a lifetime, not with collective decisions on major changes for the world as a whole. They can reveal information on what decisions individuals do make over short time horizons but not what societies *should do* over long ones.

Third, imagine that, despite even this problem with the methodology, we are so committed to the 'revealed ethics' approach that we press on. If we look historically, what we find – and this is the next problem with the imposition of a discount rate around 6% 'derived' from the markets – is that long-run, real (i.e. inflation-adjusted), low-risk rates of interest on consumption or other loans (for example, government bonds in the US or UK) are much lower, around 1.5% over fifty years, than the long-term investment returns of around 5–6%. There are numerous reasons to do with risk and imperfect capital markets that may result in different rates, including varying information or patterns of risk, the particular circumstances of investors or financiers, and so on. And if they are different, then what do we learn? Actually, the riskless real return on consumption loans – 1.5%, not the 6% that analysts often use – holds the more relevant information. This is because we are comparing paths using a measure of social welfare expressed in terms of the utility of *consumption* over the indefinite future, so that it is preferences over consumption, not returns to investment, that are at issue in specifying social values. Also, the riskless rates are more relevant because uncertainty is already directly and explicitly incorporated into the models; it should not, therefore, be separately and implicitly incorporated via the discount rate. Even so, while it has more relevance than other interest rates, this observed rate of interest of 1.5% cannot settle the issue of intergenerational values.[8]

Fourth, and again this is vital, there are many goods and services of relevance here, including, crucially, environmental goods and services. All of the discussion so far does not take into account the fact that the relative prices of different goods are likely to change strongly over the next few decades. If we invest in conventional goods and ignore climate change, we will have more conventional goods and fewer environmental goods and services. As a result, the price of environmental goods relative to conventional goods is likely to rise dramatically. This point is often missed by those who argue that, even if the damages from climate change in the far future are large, the best strategy is not to invest in reducing emissions now, but to put resources into alternative investments and use the returns to tackle climate change if and when it becomes necessary. If we do as this argument suggests and, at some stage in the future, try to 'rectify the environmental damage', 'buy the environmental goods' or 'buy down the environmental problems we have created', the cost of doing so will have risen sharply relative to the cost of acting preventatively today. We make a basic error if we think only of the discount rate for consumption goods and ignore the difference between that discount rate and the one for environmental goods.[9] In short, the 'invest elsewhere' line of argument not only misunderstands the basic theory – indeed the basic choices – when there is more than one good, but also ignores the flow–stock origins of, and the irreversible nature of, much of the damages from climate change.

There is a version of the 'invest elsewhere' argument which presents investment in climate change as an *alternative* to investing in development, saying that we would be better off devoting resources to other projects such as fighting malaria or – and the absurdity of this counterposition should be particularly evident here – promoting availability of water.[10] Of course, these are hugely important ambitions. However, the argument that we should invest here, *in preference* to climate change, makes one logical mistake after another. First, climate change and development projects interact with each other because climate change will undermine development, so to analyse them separately is an error. Second, climate change is an externality and a market failure, so rectifying the failure should lead to net gains, not net costs, for society – why would we not take the gains from correcting a market failure? Giving future generations an improved environment and less of other assets could make them better off without making the current generation worse off. Further, a strategy on climate change is not directly analogous to a road or water programme which can be analysed as a separate investment project from a given public budget. The resources to combat climate change will come out of all aspects of income, particularly consumption, and cannot therefore be seen as part of a public investment budget, as some authors try to portray when they pose the challenge of climate change as just another public investment allocation question.[11] Still further, this is about risk for society as a whole, not narrowly a single investment project, and its magnitude and relationship with other risks must be an explicit part of social decision-making. One could go on – there are multiple further layers of confusion. Such arguments are simply analytically unsound.

Overall, the logic of the route that tries to read off a social discount rate for evaluating the impacts of climate change from market rates of interest and return falls apart in at least four fundamental respects set out in terms of the 'serious analytical mistakes' described. There is no reasonable alternative to a direct examination of ethics. Information from markets and from actual decisions by policymakers has some relevance, but in using it there must be careful reference to: the nature of the decision underlying the evidence, including by whom the decision is made and over what period; the structure of markets, including their imperfections; the elements of the risk; and, finally, which goods are being examined. We then have to consider the links between all of these and the ethical decisions and values which are at issue. And we have to recognise the deepest elephant trap of them all, namely that we are choosing between very different paths and thus that discount rates depend on those choices – they do not drive those choices by themselves. All too often when discussing discounting, economists ride roughshod over the details of this logic.[12]

The costs of inaction

When considering a proposed policy to tackle climate change, it is necessary to compare the costs of that policy with, first of all, the costs of doing nothing, and, second, the costs of *alternative* proposals to tackle the problem. The (gross) costs of *action* to reduce greenhouse gas emissions were discussed in Chapter 3, which suggested that holding atmospheric CO_2e concentrations below 550 or 500 ppm is likely to cost around 1–2% of GDP for the next few decades. Now that the important role of ethics and discounting has been introduced, it is possible to consider the other side of the coin – the (gross) costs of *inaction*.[13] While earlier chapters broadly discussed the size of the damages and risks, our discussion of ethics and discounting enables us to think through in more detail how we compare damages in different circumstances and at different points in time.

Further, comparing the costs of action and inaction in this way enables us to compare costs under different policy regimes. These regimes may, for example, be two different maximum CO_2e concentration targets; alternatively, one may involve a concentration target, and the other may be business as usual with no restrictions on greenhouse gas emissions. By undertaking a range of such comparisons, it is possible to evaluate the case for a particular policy regime (for example, holding concentrations below 500 ppm) versus others.

In the remainder of this chapter, four possible approaches to comparing the costs of action and inaction under different policy regimes are discussed. These focus not only on the role of ethics and discounting, but also on the role of risk, and the scale of possible damages, in the evaluation of costs of action and inaction.

The first approach is a kind of risk analysis grounded in thought experiments. We can ask the following type of question: given the probabilities of different climate outcomes under different strategies – for example, a 50% probability of temperature increases exceeding 5°C under business as usual versus a 3% probability of the same if CO_2e concentrations are held below 500 ppm – would you pay 1–2% of GDP for a few decades to obtain the latter rather than the former? Those are questions that, I think, are transparent and easily understood. And we can similarly ask about holding CO_2e concentrations below 500 ppm versus 450 ppm, 550 ppm, 650 ppm and so on. In this example, the cost of inaction is the greatly increased probability of high temperatures with their associated severe consequences. In this case the 'thought experiment' does not require full specification of social values. All that is required is a simple ranking across the main alternatives (in the jargon of social choice theory, we need only a 'partial ordering').

To have a constructive discussion along these lines, we require guidance on the probabilities of the increases in temperature, and some understanding of the consequences. The science has given us very useful guidance on both these issues. There is now a sufficient scientific basis for us to make decisions, even though there is much more work to be done. Most importantly, the science tells us that action is urgent: the way in which flows accumulate into stocks, together with the difficulty of reversing stock increases, imply that delay is dangerous.

This same logic indicates the dangers of the 'slow policy ramp' – in other words, weak action now followed by stronger action in the future – advocated, for example, by William Nordhaus. As we saw from Chapters 2 and 3, this would, within a few decades, put us in a position where it would be difficult to avoid reaching 600 ppm or 650 ppm CO_2e (we would be at 550 ppm CO_2e by mid-century). At 650 ppm CO_2e, the probability of temperature increases exceeding $4\,^{\circ}C$ would be 58%; of exceeding $5\,^{\circ}C$, 24%. That is a very dangerous place to be. In conducting a risk analysis of the consequences of such potential temperature increases, most would judge that the costs of inaction are greatly in excess of the costs of action.

A second, related, way to think about the costs of inaction in relation to the costs of action under different policy regimes is in terms of prospects for growth and development. Unchecked climate change will, within fifty or a hundred years, make it very difficult to achieve continued economic growth in both the developed and the currently developing world. Climate change is likely to choke off the prospects for growth. Thus we can ask whether we wish to have slightly higher incomes for a few decades, perhaps just 1 or 2% higher, or to forgo these slightly higher incomes in the short term in exchange for a much higher growth rate in the hundred years or so which follow. Moreover, by taking strong action we might start a wave of discovery which could increase growth rates in the near future.

A third, less formal way to think about the costs of inaction versus the costs of action under different policy regimes is to think about the nature of the world and the quality of life which would arise with inaction compared to that with strong action. The latter involves a more collaborative, cleaner, quieter, more biodiverse and safer way of living, including growing consumption of goods and services. The former involves growing pollution, continuing waste, destruction of forests and biodiversity, and increasing competition and quarrelling over ever more scarce hydrocarbons. It is extremely difficult to be quantitative about these types of comparisons, but the issues involved can be understood by anyone who thinks carefully about the different implications of different strategies.

Finally, and without wanting to detract from the crucial insights that can be obtained from the first three perspectives, quantification of key impacts from climate change is possible via formal modelling. This is generally undertaken via the Integrated Assessment Models discussed earlier in this chapter. Going down this route involves trying to estimate explicitly the costs of damages avoided by a given strategy relative to another.

What are the likely costs of inaction with which we should compare the costs of action outlined in Chapter 3 (1–2% of GDP for holding concentrations below 550 or 500 ppm CO_2e)? The action with which we are concerned will involve holding concentrations below a given level. In the modelling for the Stern Review, we calculated the welfare difference between 'business as usual' and 'no climate change' and calculated the welfare difference between 'holding below' a given concentration and 'no climate change'. The comparison between the costs of inaction on the one hand, and action to hold below the given concentration on the other, is then given by the difference between the two calculations. The costs of business as usual are so large that the first calculation overwhelms the second for maximum concentrations of 500 or 550 ppm CO_2e. These models are too crude in their damage assessments for the more subtle choices *between* 550, 500 and 450 ppm as maximum concentration levels.

To pursue this fourth approach to the costs of inaction we need a model that encompasses, and makes assumptions on, all five of the key links in the climate change process: economic activity to emissions; emissions to stocks or concentrations; concentrations to temperatures; temperatures to climate change; and climate change to effects on welfare. We also have to introduce assessment criteria, or 'objective functions', for comparing two states of society; these will be based on the kind of philosophical judgements discussed in the section above on ethics and discounting. Thus we need structural *and* ethical assumptions. Both are very important in shaping numerical results.

The aggregate modelling of the Stern Review was presented in just one of the thirteen chapters on mechanisms, impacts and targets that made up the first half of the review. In the other twelve chapters the probabilistic approach to outcomes and impacts were set out in some detail, and the costs of different kinds of action were examined. Nevertheless, Chapter 6 attracted a lot of attention because it provided an aggregate statement of costs of action versus costs of inaction.

While this approach is described as 'aggregate modelling' because the models operate at a high level of aggregation, it is important *not* to think of such models as providing an 'adding up' of country-specific and sector-specific effects. Rather, they are highly simplified macro-level models that force many complex phenomena into just a few variables. For example, the damages from a given temperature increase are calculated not by adding up the damages from the possible submergence of parts of Bangladesh, the desertification of parts of South-West Africa or the conflict arising from mass migrations, but rather by a very simple formula relating damages to temperature increases.[14]

We found, using a structural model with risk at its core, and calibrating damages – or costs of inaction – in terms of recurring losses to GDP, that the damages from business as usual were, in welfare terms, 'worth' between 5–20% of GDP per annum. Not all of these damages would be avoided if we were to set a ceiling of 550 ppm CO_2e, but many of them – certainly enough to justify the expenditure of 1% of GDP – would be.[15]

The estimates arising from this modelling in the Stern Review, and from the base case in particular, have been widely quoted as saying that the costs of inaction (or benefits of action) are much greater than the costs of action. They do indeed give a quantitative expression to the judgement that it is well worth paying for the costs of rapidly reducing emissions and thereby reducing the probabilities of very damaging outcomes. Thus, as a simple way of communicating, this formal modelling-based approach had and continues to have some usefulness. These models also allow one to illustrate the effects of changing some of the underlying structural or ethical assumptions on damage estimates. It is crucial to emphasise, however, that the precise quantitative results of such models should not be taken too seriously. They are very sensitive to assumptions and leave out altogether crucial details such as conflict and different ethical approaches, and are usually weak on their treatment of other issues such as risk or biodiversity. As the modelling section of the review concluded, 'we therefore urge the reader to avoid an over-literal interpretation of these results. Nevertheless, we think that they illustrate a very important point: the risks involved in a "business as usual" approach to climate change are very large.'

Looking back, I think the Stern Review assumptions led to an under-estimation of the costs of inaction *whichever* of these four approaches are adopted. Current evidence is now showing that the review was too cautious on the growth of emissions, on the deteriorating absorptive capacity of the planet, and on the pace and severity of the impacts of climate change.

Why some economists got it so badly wrong

Given the potentially immense consequences of no, weak or delayed action, I am often asked why it is that prominent economists got things so badly wrong in arguing that, overall, discounted costs of inaction or weak action were low. Their arguments on policy, such as the 'slow ramp', would lead to concentration levels so high that most reasonable people would instinctively and fundamentally disagree with their conclusions. That should, at the very least, have led them to revisit their models and assumptions; when we do, the basic flaws in assumptions and method become clear.

These authors[16] made assumptions on the structure of their models which generated damages that were totally implausible – such as mean damages from a 5°C temperature increase of around 2% of GDP[17] – and they provided only minimal treatment of risk, which is at the heart of the issue. In addition, as we have argued, their assumptions and discussions of the ethics, especially their argument that ethical values are revealed in markets, fly in the face of the basic economic theory of policy in risky and imperfect economies; in relation to discounting in particular, they make one mistake after another.[18] Behind all the technical complexities of the modelling were two simple assumptions: first, that the risks are small; and second, that benefits in the far future matter very little. While you can cloak these in technicalities, it is clear that if you make these two assumptions then you will conclude that the costs of inaction are likely to be small.

It is not easy to understand why these mistakes were made. One reason could be that they missed the scale of risks indicated by the science – the scientific evidence has become clearer and more worrying over time. Another could be that the technical treatment of large risks in these models is not straightforward computationally and analytically,[19] and that they took the 'easy way out' in their modelling by ignoring major risks and assuming that others were implausibly small.

Academics, like many others, tend to hang on to and defend positions they have taken in the past, notwithstanding that scientific evidence moves on and that the analytical errors in their methods start to emerge. Happily, the balance of the argument in the economics profession is changing as the key issues are examined more closely.

A broader and more fundamental reason is that many of the authors forgot, overlooked, or never knew the basic necessary underlying theory of policy, particularly that of policymaking in risky, imperfect, economies with many goods (in contrast to standard macroeconomics with its focus on GDP). 'Imperfect' here means that there are market distortions to do with taxation, rigid prices, limited information and the like. This theory was extensively applied to project appraisal and planning in the 1960s and 70s and follows in the tradition of James Meade, the famous Nobel prizewinning economist of the London School of Economics and Cambridge University. It was central to Meadean theory, for example, that there were many discount rates depending on which good, which person or which circumstance was being examined, and that the relevant set of discount rates and factors depended on the development path under consideration. Risk, market imperfections and multiple goods are all of fundamental importance in the context of climate change, and the mistakes from being unaware of or ignoring this literature are illustrated in the discussion of the logical dangers of using market rates of interest or return as social discount rates.

Why have many economists forgotten or ignored this theory, given its basic relevance? Doubtless the answers vary across authors. Some who are narrow modellers never learned it. This type of theory of applied policy lamentably began dropping away from graduate courses in the 1980s.[20] Others are reluctant to take on the ethical discussion that policymaking for climate change will inevitably involve. The ethics are seen as something that economics can and should avoid. This is, again, reflected in the way economics has come to be taught in many graduate schools. The absence of this type of Meadean theory of public policy in graduate courses may also be related to the zeitgeist of the 1980s and 90s with its presumption that governments were the problem rather than the makers of sound policy. One could speculate further, but the fact remains – the relevant theory dropped far down the list of 'what economists should know'.

I trust that seeing the mistakes that arise when the economics of policy in economies characterised by risk, market imperfections and multiple goods is forgotten or ignored will lead to its restoration at the centre of graduate courses. That is where it should be. The damage is not confined to the economics of climate change. The recent turmoil in financial markets is a further reminder that the standard model of perfect competition in perfect markets cannot be a sensible way to see the role of policy in the world we inhabit. That is not to say that we should abandon markets. On the contrary, it is to say that we should make sure we apply the relevant theory by thinking, when making policy, of how to remedy, or take account of, their imperfections.

The challenge of implementation

This chapter has discussed four ways in which one might, by considering the role of ethics, discounting and risk, compare the costs of action and inaction under different policy regimes and, in so doing, arrive at an overall conclusion as to the desirability of a given policy. I have argued that the case for action to hold CO_2e concentrations below 500 ppm – indeed, for strong and decisive action to this end – is very powerful whichever of the four approaches one adopts. The time has come to examine how we can get to a point where concentrations of CO_2e are held at or below that level. This will bring in a whole range of economics as it relates to development, growth, international trade, risk, finance, industrial organisation, technical progress, game theory, and so on, as well as the obvious subjects of public and environmental economics. With such all-embracing subject matter, we have to use all the tools of the profession, and develop our analytical techniques still further.

Obviously science and technology are central to the practical challenge, but there are also important issues of international relations, law, geography, history, sociology and anthropology. It is impossible for one person to handle all of these at once, so it is very important that economists be aware of the importance of the insights from other areas, and that research be organised so that disciplines interact and collaborate.

In the next part of the book, the focus is mainly on the principles and practice of the relevant economic policies and on how to build, support and sustain international collaboration. This will involve what many of us refer to as political economy, the subject of which is political decision-making where people are assumed to understand their own interests – it involves the examination of how that understanding influences decisions and the way they are implemented.

There will be dislocation in the transition to a low-carbon economy, one that in its sources and use of energy is very different from the economy that exists now, even though our activities may not be so different.[21] Dislocation generates opposition. There will be groups which for understandable reasons oppose change. How to manage this adjustment is a crucial part of the challenge of climate change policy.

All of this requires public discussion of a nature and quality which is rare. It requires leadership and international collaboration on a scale which is unprecedented. We can see the basics of the investments, technologies, economic policies and international structures that are necessary.

Can we get there?

CHAPTER 6

Policies to reduce emissions

The challenge is to produce an effective, efficient and equitable set of principles and policies to guide both national action and a global deal. If the global deal is not effective, we will be sentencing future generations to living in a very risky world; if it is not efficient, we will have wasted resources and possibly undermined support for action; and if it is not equitable, we will not only be treating poorer people unjustly, but we also risk damaging the international coalition for action which is vital for success.

The economic and technological instruments that are adopted and the arrangements for international action that are made – in other words, the global deal – have to reflect these three criteria. While it was the economics of risk and sensible insurance that pointed us in the direction of reducing emissions via quantity targets, it is the economics of cost that points us to the importance of market-related mechanisms, and a price for greenhouse gases, as the best ways to promote the search for the cheapest ways of achieving those emission reduction targets. It is also important to look at problems with related markets that could hinder action. An understanding of what is more or less equitable needs an explicit discussion of equity criteria within and across nations.

If public actions and policies give the right signals and rewards for cutting greenhouse gases, then markets and entrepreneurship will drive the response. The bulk of the action will be in the private sector – this is not about a return to governmental control and rigid planning; on the contrary, it is about enabling markets and private-sector initiative to work well.

We do not, in this chapter, discuss the challenges involved in reducing emissions from deforestation, the source of around 20% of emissions;

some of the principles described in this chapter apply, such as the importance of trade, of seeking out lowest-cost reduction options, and of integration into a strategy for growth, but other issues are very specific to deforestation as well as to international collaboration. These are discussed in Chapter 8.

Prices and markets

Why should we want prices to be at the heart of policy? There are, broadly speaking, two reasons. First, the right prices can give an incentive for consumers and producers to do something, i.e. cut emissions; thus we 'correct the externality' by embodying in the price of a good not only the cost of the raw materials, labour, capital and so on used in its production, but also the cost of the damages from the emissions produced in the consumption or production of that good. From this perspective, the price of carbon should reflect the marginal social cost (MSC) of greenhouse gases – the cost to society of emitting one extra unit of emissions. This is the economic expression of the idea that 'the polluter should pay'.

The second role for prices is to achieve reductions of greenhouse gases as efficiently as possible – in other words, to keep the costs of action down. By having just one price for emissions we ensure, at least in principle, that all opportunities for reducing emissions which cost less than the price will be exploited. As long as all these opportunities are exploited in any given area, then the last (or marginal) opportunity to be selected will have a marginal cost of reduction (the marginal abatement cost, or MAC) just equal to the price of greenhouse gases. If the same price applies to all areas – and if markets work well – the marginal cost of cutting emissions will be the same everywhere. In other words, we will not be able to reduce the cost of abatement by shifting from one option for emissions reduction with a higher marginal cost to another option; we will, therefore, have cut emissions efficiently. It is, of course, a big 'if' to assume that markets work well everywhere; the possibility that they may not forms, or should form, a major part of policy analysis.

How do we work out the right price? In a simple world, perfect in all relevant economic respects other than this greenhouse gas externality, then a feature of the 'right' level of abatement would be that the MSC (the damage from an extra unit of emissions) and the MAC are equal. If the former were larger, for example, then a net social gain could be attained by reducing emissions further (the gains from reduction – the MSC – would outweigh the cost – the MAC). But the world is far from perfect in all relevant economic respects. Imperfect information, risk, the absence of well-functioning and comprehensive insurance markets, unrepresented consumers (future generations), profound inequalities, and imperfect or absent capital markets are all important considerations when it comes to making policy on climate change. That is why we cannot base policy only on the simple-minded approach of setting a tax equal to the estimated marginal damage from greenhouse gases and leaving the rest to the markets.

Those responsible for appraising and assessing policy should be constantly checking that the MACs are similar across all areas of emissions reductions, and whether they are roughly similar to assessments of the MSC of emissions. However, this last check, while necessary and important, is likely to be fairly imprecise because of the very large possible range for the MSC, often called the social cost of carbon (SCC).

The SCC is calculated by estimating the damages created by the emissions of an extra unit of carbon, keeping in mind that this extra unit results in higher concentrations of greenhouse gases in the atmosphere over the very long term. That calculation depends on the assumptions made on the future path of the economy and of emissions, the strength of the carbon cycle shaping absorption, climate sensitivity, distributional and intertemporal values, and so on. If you also factor in assumptions about probability distributions of all those things and attitudes to risk, you can get a huge range of estimates for the SCC. This means that the SCC is a very weak and unreliable peg for policy. That is why I think it makes for a more sound and transparent policy analysis to avoid starting with an attempt to work out the level at which the MSC and MAC are equalised, and instead to follow an alternative route as outlined in the first part of the book. This alternative considers the appropriate targets from the perspective of risk and costs, and seeks out the cheapest method, generally using a price mechanism, of reaching the targets. The MAC associated with that method becomes our guide price for greenhouse gases and should be checked to see if it lies within the very broad range of possible MSCs. To start with an attempt to calculate strategy *directly* from an equality of the MSC and MAC is likely to be arbitrary, unreliable and inconclusive because of the former's great sensitivity to underlying assumptions. The principle of the equality of MSC and MAC does provide a sound guide for policy: the issue is how it is best applied in an uncertain and risky world.

There are, broadly speaking, three policy instruments that can be used to put a price on greenhouse gas emissions: a tax, in which case the price is determined by the level of the tax; trading of a fixed number of emissions permits, the number and initial allocation of which is set by the government – in which case the price is set by the interaction of buyers and sellers in the market; and regulations or technical requirements which may need more costly equipment or processes – the implicit price is then the extra cost divided by emissions saved.

Each of these has its advantages and disadvantages. There will be some circumstances where one will be preferable to the others; given the different pros and cons, the balance of policies might vary across places and times.

All three approaches require the measurement of carbon emitted, but there are some circumstances where the burdens of measurement differ across these tools. Taxes on carbon sources such as hydrocarbon fuels could be set, reasonably accurately, in relation to the amount of emissions their use is likely to generate. In these cases the administrative burden of taxation could be quite low, as it is with petroleum products. A tax or licence on a vehicle is much less precise since it takes no account of how much the vehicle is used. It has its role to play, however, in focusing minds when purchasing long-lasting equipment, as we know that many decision-makers, particularly consumers, do not always think through the consequences of their decisions when they stretch out over time.[1] Taxes on carbon sources also have advantages when there are so many small users that detailed individual measurement of emissions is difficult. In such circumstances, there are advantages in levying taxes early in the production chain.

Taxes, however, have three important disadvantages of special relevance in this context. First, they do not give us much certainty on how big the resulting reductions will be. Estimates of responses to taxes are imprecise and many of the effects operate with long lags. It may be several years before we discover that a tax is not quite as effective as we hoped it would be in reducing demand; by then, substantial emissions likely to affect stocks in the atmosphere for centuries will have taken place. Second, taxes are very hard to coordinate internationally as countries are very protective of their tax independence. Third, electorates are mistrustful of governments' use of tax revenues. There is an ever-present assumption that this is just another excuse for taxation, or that it is a 'stealth tax'. They may not believe that taxes elsewhere, or borrowing, will be reduced, or that revenues will be spent well.

Sometimes earmarking can provide some assurance on how revenues will be spent – some revenues from greenhouse gas taxation could be directed, for example, to research and development on green technologies, to reduce 'fuel poverty', or to fund adaptation in developing countries. Of course, detailing where revenues are to be spent does not guarantee that they will be spent well – this is another and very important challenge – but earmarking may help to overcome political opposition. Many economists have traditionally been hostile to earmarking as an unnecessary and possibly wasteful constraint on government's ability to direct expenditure to its best use, but the politics of setting taxes are often not as simple as the economics.

Stéphane Dion, former leader of the Liberal Party in Canada, promoted a 'Tax what you burn, not what you earn' approach.[2] This is a simple, pithy statement which tries to indicate to people how revenues from 'green taxes' will be used (in this case, reducing income tax). We must be careful, however, not to exaggerate the potential size of this source of revenue. Work for the 2008 Mirrlees Report on Taxation in the UK suggests that 'green tax' revenues are unlikely to exceed 1% of GDP – around £15 billion per annum – which is also my own ballpark estimate. Adjustments for necessary public expenditures in this area take the sum down even further. This is not enough to finance major cuts in, say, income tax or VAT. On the other hand, it could make a very big difference to the UK's contribution to world research and development on energy and the environment, to halting deforestation and to funding economic development in a more hostile climate.

Carbon trading on the basis of quotas can, in some respects, be understood as the 'opposite' of carbon taxation. In the former, the government sets the quantity of emissions; given this quantity, the market then determines the price of emissions. In the latter, the government sets the price of emissions; given this price, the market then determines the quantity of emissions. Because of this dualism, the advantages of quotas are related to the disadvantages of taxation, and the disadvantages of quotas are related to the advantages of taxation. One advantage of quotas is that they give greater certainty about the quantity of emissions, because that quantity is directly set by the government. When it comes to greenhouse gas emissions, quantity certainty is of real importance because, as we showed in Chapter 2, the risks associated with exceeding a given concentration target are severe. The flipside, however, is price uncertainty, on which taxes bring greater clarity. We cannot have both price and quantity certainty in an uncertain world.

Another advantage of quotas and trading is that they allow inter-national private-sector flows of carbon finance from rich to poor countries. This will be of great importance in securing a global deal. Quotas may be less attractive administratively from the point of view of measurement for industries with many small firms, but can work well in industries where installations or firms are bigger. The European Union Emissions Trading Scheme (EU ETS), the largest trading scheme for greenhouse gases, covers more than 40% of the EU's emissions, despite including just a few major kinds of economic activity: power and heat generation, and selected energy-intensive industrial sectors such as combustion plants, oil refineries, coke ovens, iron and steel plants, and factories producing cement, glass, lime, bricks, ceramics, pulp and paper.

There is a further argument for quotas, which is of substantial relevance here. This is that there are powerful oligopoly elements in oil and gas prices – these are not markets where we can be confident that (untaxed) prices reflect marginal costs. We do not know how an internationally coordinated tax would affect world hydrocarbon prices. This unpredictability of tax effects is a strong argument for quotas in a context where missing targets can be very damaging.

Regulations and technical requirements have their advantages where other markets might work problematically or slowly, or where economies of scale and technological predictability are important. Standards designed for purposes other than combating global warm-ing, such as those requiring catalytic converters on cars to reduce smog, or unleaded petrol to reduce damage to child development, work effectively and probably more cheaply than would special taxes on the relevant pollutant. Where car-makers know these standards are likely to be global and long term, all cars are designed to meet them – costs reduce rapidly and results come quickly. And they are often much cheaper than the industry predicts: in the vehicle-emission control programmes of the Environmental Protection Agency in the USA, for example, industry stakeholders predicted price changes to consumers that exceeded actual changes by ratios ranging from 2:1 to 6:1.[3]

The right mix of these types of policy requires careful examination of particular circumstances. In the UK we are likely to use all three of taxes, quotas-cum-trading and regulation, as is much of the world. Europe has strong taxes on petrol as well as the Emissions Trading Scheme, and maintains tight regulation on many pieces of equipment. So too does the USA, although the balance is very different and trading schemes are at present only local and mostly in the early stages.

There is no problem with countries having different combinations of policies, but there are two important provisos. First, the overall level of ambition should be strong and equitable. Second, providing a strong role for trading schemes is important to allow international trade in greenhouse gas reduction, which will both improve efficiency and provide incentives for developing countries to join in international action. These 'carbon flows' will be a key element in the 'glue' that holds the global deal together.

These principles can provide very practical guides, but there is much more to policy, including the operation of trading, and other market failures. Before we get to these issues I want to examine briefly three important questions concerning or influencing prices and price structures: the price level likely to be necessary; the clarity and stability of signals; and income distribution.

First, as we have seen, the simplest and most direct way to look at the appropriate price of greenhouse gases is through the cost of abatement and, in particular, the MAC curves of the type produced by McKinsey and illustrated in Chapter 3. To hold concentrations below 500 ppm CO_2e successfully, an approximate 30 Gt cut in annual flows is necessary by 2030. In a well-functioning market this would require a price of around €40 per tonne of CO_2e on the basis of the MAC interpretation of the McKinsey curve. The current price on the EU ETS (summer 2008), near the beginning of Phase 2 of the scheme, covering the period 2008–12, is around €25. As ambitions are tightened and we move into Phase 3 (2012–20), a price in the region of, or higher than, €40 per tonne is quite likely.

A key number to look for in determining the price is the economic viability of carbon capture and storage for coal. While some renewables might soon be competitive with hydrocarbons even without a carbon price, the scale of coal use and the challenge of providing power in many countries (including China and India where coal is available, cheap and secure) means that CCS for coal will be necessary, at least for the next few decades. Electricity generated using CCS for coal will always be more costly than electricity generated from coal without CCS, and the viability of CCS requires a carbon price that covers these extra costs. For 2030, McKinsey's current estimate of the necessary price to sustain the installation of CCS retrofitted to existing plants is in the range of €30 per tonne CO_2e, and for new CCS around €20 per tonne. Clearly these estimates will be revised with experience, which we need to get as soon as possible. The additional cost of CCS in commercial-scale prototypes is likely to be much higher than this. Because many results from learning are available to all, and because the technology is potentially so important, strong public support for these first commercial-scale plants is vital. Without it we are unlikely to see progress. It is vital that demonstration plants on a commercial scale are in operation within ten years; we will then know much more about the necessary price. Further, any detailed analysis must take account of the historical structure of the capital stock and the economic lifetime of equipment under different assumptions on prices.

Second, many investment programmes where emissions are of great importance – including those for electric power, production of vehicles, and design and construction of buildings – are large, have long gestation periods, and last for several decades or more. This implies that the price signals and regulatory structures should be clear, extend over the long term and be predictable (what some people call 'loud, long and legal'). That is why agreement across political parties within a country is of great value, as are long-term treaties across countries. At a regional level, for example, European structures can, in some circumstances, give additional clarity and confidence to national policies. Having strong ambitions embedded in Phase 3 of the EU ETS – a 30% cut in emissions by 2020 in the context of an international agreement – would be a powerful and clear signal.

Third, putting a price on carbon emissions will raise the cost to the consumer of a range of goods and services – heating and travel, for example. In the short and medium term, this will affect poorer people most. Low-income groups can be protected via direct transfers, through the tax and transfer system. Indeed, this would be an important call on extra revenues from 'green taxes' or their equivalent. There are also other methods, including 'lifeline' tariffs, which allow a small element of electricity supply, for example, at lower prices. We have the instruments to protect poorer people and we should use them.

Trading schemes

Trading schemes work by allocating rights to emit greenhouse gases and then allowing trade in the certificate containing the right to emit. If a company emits less than the total of rights to emit that it holds in certificates, it can sell its excess certificates; if a company plans to emit more, it must buy additional certificates. Such schemes therefore require allocation plans for certificates, monitoring of emissions and arrangements for markets for trading. Pollution markets also require clear and consistent price signals to function smoothly. In the US market for tradable sulphur dioxide allowances, which originated in the late 1980s, it took several years for prices to stabilise, with the price for an avoided tonne of SO_2 fluctuating from lows of $70 to highs of $1,550. Over time, players became accustomed to the system, policy rules became clearer and financial intermediaries entered the picture, all of which contributed to greater price transparency. Once a degree of certainty emerged, the SO_2 markets took off, with emissions declining strongly over time.

There are many intricacies of trading schemes which are vital to their successful operation; again, details matter. In a short and wide-ranging book we cannot go into all of them. I want to focus on just three crucial and related issues: the allocation of allowances; the predictability of prices; and openness to external trade.

The allocations of allowances in the EU ETS were initially almost entirely free of charge and based largely on past history and articulated existing plans. They were made at the national level in a National Allocation Plan, and then examined by the European Commission for overall consistency with EU policies, particularly in relation to the question 'Do they add up to the target reductions?' In Phase 1 (2005–7), prices started at €10–25 per tonne of CO_2e, but there was a sharp fall in April 2006, when it was revealed that larger amounts had been allocated than originally envisaged. Many cynical observers have used this occurrence to argue that emissions markets do not work. The European Commission was, however, much stricter with allocations for Phase 2, leading to the 2008 price of around €25 per tonne.

There are simple and clear lessons to be learned from this experience. Total allocations and the cuts they embody must be transparent and clear. And if prices are to drive substantial reductions, then these reductions must be embodied in the allocations. In other words, if allocations of rights to emit are too high, then reductions in emissions and prices will both be too low – pretty basic economics. There is no alternative to tough allocations, and 'special pleading' or 'gaming' must be resisted. (Gaming can arise, for example, in 'grand-fathering' allocations, whereby firms can overestimate past emissions, future plans or difficulties in adjustment, and gain substantial profits from large allowances.)[4]

Over time, the best way to avoid over-allocation and gaming is to auction all allowances. Any rational firm would obviously rather have free allowances than be forced to pay for them, but there are major problems with arguments that firms should have the right to pollute for free up to a certain level. The argument for *temporarily* free allocations, or for phasing in the auctioning of permits – that they help with the adjustment process, given that original investments and commitments were made in a context without carbon prices – has more substance.

In the long run, however, auctioning is superior to free allocations in three crucial respects. First, it raises revenue for the government. Giving away that revenue as transfers to firms through free allowances would be a peculiar and inegalitarian use of public money relative to, say, supporting poor old-age pensioners or the disabled, or health and education services. Second, auctioning can hasten adjustment. The longer allocations are given free, the less pressure there is on firms to move quickly. It is true that the marginal incentive of a carbon price should give a strong reason to economise on emissions even if allocations are given free. But that pressure is intensified if weak or no adjustment implies significant losses, rather than profits simply being lower than they might otherwise be. It is likely that shareholders and capital markets would react differently, particularly since losses are more identifiable than the possibility of higher profits. Third, the process of 'grandfathering' free allocations gives special privileges to incumbent firms and thus undermines competition by disadvantaging potential entrants into affected industries.

Predictability of prices is important to investment planning, so reduction targets and allocation mechanisms should be set over extended periods. The range of 2012–20 for Phase 3 of the EU ETS seems to be about right, especially if plans for subsequent phases are made early. There will inevitably be some price fluctuations because quantity reductions are fixed and economic circumstances are uncertain, but they will be smoother if there are more margins of sub-stitution, whether across industries and regions or over time. 'Banking and borrowing' of reductions over time, for example – doing more now if attractive reduction opportunities arise, or doing more tomor-row if reductions today look particularly difficult – is not only efficient but also leads to more stable prices.

The importance of confidence in future prices has led some to suggest 'floors' and 'ceilings' for prices. For this to happen, some entity, presumably public, has to be ready to be a seller at the ceiling and a buyer at the floor. Ceilings, for example, could come with related effects such as long-term government contracts to sell reductions at a set price in the future. The motivation for having floors and ceilings is understandable, but if they do come in, they should do so only after more market-oriented and efficient structures – such as expanding the markets across activities, time and regions – have been introduced. The broadening and deepening of markets is likely to be a better route to price stability in terms of both efficiency and of achieving the necessary reductions. The flip side of achieving greater price stability via price controls is that they also lead to *less* quantity certainty – in other words, price controls put the achievement of reduction targets at risk. Another consideration that weighs against the introduction of price floors and ceilings is that it interferes with mechanisms for banking of certificates – if certificates are bought at an artificially low price, it should not be possible to save them for use at a later date.

Opening up trade across regions is also of great importance in the sense that private-financing flows from carbon markets will bring some developing countries into a global deal. The EU ETS is likely to be the benchmark for defining relationships between countries and regions. Schemes will probably emerge before long in Australia, New Zealand and parts of Asia, while the north-east US has just begun a trading initiative for electricity, and federal structures are likely to be decided in 2009 or 2010 with the new administration.[5] As trading structures in these rich countries develop, it will be vital to build consistency between regions, and for overall targets to be agreed. A region that has ambitious targets would understandably be reluctant to agree to trade with an area with weak targets, since such collaboration would water down global ambition. Floors and ceilings in region A would make interaction with region B more complex – if a price is kept low in region A, for example, all those in region B would want to buy at the lower price and trade would have to be restricted.

Trading between developed and developing countries is crucial, but has to be conducted in two steps. The first would involve one-sided trading building on the ideas, but with different structures, of the current Clean Development Mechanism (CDM). The notion of one-sidedness means that an enterprise in a developing country can get credits for reducing emissions relative to a benchmark, but is not penalised for emissions above some level. The CDM benchmark is defined in relation to what the enterprise might plausibly have done – continued with past practices, for instance. It is subject to an 'approved list' for the new technologies – examples of which include the building of hydropower plants in South Africa by UK project developers, or the conversion of coal-mine methane to electricity in China financed by a German investment bank – and requires the support of specific in-country committees set up by the national authorities of each developing country for this purpose. Projects are then approved by a secretariat in Bonn. The CDM has established the principle of one-sided trading and has started to build significant interest in low-carbon options in the developing world, with sizeable funds now beginning to flow.

However, the current CDM institutional system could not be very easily expanded to the necessary scale. Its definitional structure, working at the level of the firm and relying on the 'counter-factual' of what would otherwise have been done, is just too cumbersome for administration on a large scale. The expansion of one-sided trading that is needed if developing countries are to play, as they must, a strong role, requires wholesale, and greatly simplified, trading. For example, a city could put in place an overall plan for reduction in emissions from transport, and then trade the reductions achieved. Or a province of a country could set up a CCS programme for coal and sell the reductions from the whole programme.

By 2020, we should aim to have commenced the second stage of integration of developing countries into world carbon markets – namely, targets and an integrated two-sided trading scheme for *all* countries.

Promoting technology

There is a whole swathe of important market failures and related issues which require us to go beyond simply a price for carbon if policy is to work: entrepreneurs, households and decision-makers cannot know with certainty what the price of greenhouse gas emissions will be over the indefinite future; there are issues of credibility, gaming and asymmetric information since we cannot rely on future governments' determination to set good policies; it is difficult to be absolutely sure that we can chart future growth paths and thus associated relative prices and costs, and this can amplify market failures associated with future policy; and the development of technological ideas is a process that itself generates multiple externalities (or spillover effects) which must be taken into account. A number of market failures and uncertainties interact.

Many commentators are sceptical about technology policy, saying it is wrong for bureaucrats to 'pick winners'. There is something in this, but it is also naive, or dogmatic, in its underlying assumptions that markets work perfectly unless distorted by government. In this case, markets do not work well unless *assisted* by government.

The first reason for having technology policy in addition to price policy – and it is a general argument for supporting research and development (R&D) – is that an idea is essentially a public commodity. Its use by one person does not prevent others from also using it – in other words, the additional cost of its further use is zero. If this were the only consideration, basic economic reasoning would suggest that ideas should be freely available. The problem is that people are discouraged from developing an idea if there is little market reward, because others instantly make use of it. This is of course why we have patents – drugs companies, for example, require confidence that they will be protected from cheap copies, at least for some time, in order to recoup the costs of R&D.

A similar argument applies to the ideas that come from experience. As techniques are used over time we become better at them – in other words, there is learning-by-doing. And there is also learning-by-watching as we absorb the experience of others. In some of the newer ways of producing electricity, such as solar and wind, better techniques have developed very rapidly from experience, bringing costs down substantially – one company's use of these technologies benefits other companies in the industry, because their production activities can help to 'push out' the technological frontier.

The importance of creating ideas and learning from their use, both of which benefit others, indicates that there is a gain arising from the actions of those who create, develop and deploy these ideas which an uncorrected market would not reward, in which case these ideas would be undersupplied and underused. The basic argument for public support for R&D, and for the rapid deployment of technologies, is especially strong in this area, not only because learning-by-experience is important here, but also because, given the dangers we face, we need ideas quickly. The arguments are stronger for newer technologies and for technologies that are furthest from the market. It is here that capital markets, given the risks, may function particularly weakly.

There are a number of possible ways to encourage the rapid development of emission reduction technologies. Those that involve basic R&D are likely to be best oriented through institutions such as universities and research establishments. Academic staff need time to pursue research, support needs to be given to programmes focused on the most promising areas, and the transfer of knowledge from research institutions to practical applications needs to be funded. In all these areas, good practice allows and encourages the combination of public and private resources. One example of a fundamental research priority for climate change policy is energy storage, which encompasses research into different kinds of batteries (including nanobatteries), into storage through the creation and containment of hydrogen, into storage through lifting or heating of water, and so on. Other priorities include photovoltaics and new materials, new biofuels, enhanced photosynthesis and nuclear fusion. There will be many more. One which may turn out to be interesting is the transformation of carbon dioxide into solids which could then be used as construction material, or for surfacing roads. This would obviate the need for underground storage of CO_2, once the technology to capture the gas has been developed. It is of vital importance that research institutions around the world be supported to pursue new ideas in an open way. The interest is intense and there is strong enthusiasm from the private sector to be part of public-private partnerships, as well as to pursue their own ideas with their own resources and structures.

Given these priorities, the worldwide decline, until recently, of public spending on energy R&D is both striking and worrying. In the two decades from the end of the 1970s, spending roughly halved in most rich countries from around $20 billion to $10 billion.[6] As a percentage of GDP, it fell from about 0.15% in the UK and USA to 0.03% in the US and 0.01% in the UK.[7] There were a number of factors at work here including disillusionment with nuclear power (dropping from around 80% to 40% of public energy R&D budgets worldwide),[8] falling oil and gas prices, and the privatisation of many energy industries. Globally, private energy R&D has also been, until recently, on a strong downward trend, roughly halving from around $8–9 billion in the early 1990s to close to $4 billion in the early 2000s.[9]

These low and falling expenditures on both public and private energy R&D contrast markedly with worldwide energy subsidies of around $250 billion in 2005.[10] These subsidies should be cut globally. They are wasteful and distorting, and it is usually the better off who benefit.

Reversing the decline in both public and private R&D is an urgent priority. It is not easy to provide a target since it takes time to expand R&D and the returns are difficult to measure. Given the declines, a target of 0.1% of GDP in OECD countries seems realistic and relatively modest in relation to the urgency of the challenge we face. In today's prices that would deliver something in the region of an additional $35 billion per annum.[11]

Of course, R&D expenditure must be used wisely. It is not just a matter of throwing money at the problem. There must be a balancing of breadth and freedom in the development of ideas with a sense of purpose in focusing on the biggest challenges to, and best opportunities for, emissions reductions. While fundamental work on photovoltaics, enhanced photosynthesis and nuclear fusion, for example, is likely to be heavily based in universities and research institutions, CCS is likely to be driven by electricity companies working with governments.

International collaboration will be crucial in areas such as fusion, while in others, competition will be a powerful source of research delivery. Some cross-country specialisation is likely to arise, and as long as we share results all these processes should be encouraged. If, for example, Japan becomes good at photovoltaics, Germany at wind and France at nuclear, we can all benefit.

Accelerating the development and use of new technologies is not just a matter for R&D – demonstration and deployment are both of great importance. There is a powerful argument for deployment support – in other words, support for the wide-scale deployment of relatively 'young' technologies that have been shown to work and for which strong learning-by-doing is likely. Such support should be firmer the more youthful is the technology, and the greater is the learning potential, subject, of course, to the judgement that it has a good chance of becoming competitive at a reasonable carbon price. Feed-in tariffs for types of electricity – providing a guarantee that one will pay a certain price for a particular type of electricity, whether it be wind, solar, or renewables generally – has been one important and effective way of building new types of capacity. The value to investors of a feed-in tariff is that a price is fixed for buying in electricity from a particular source over a long period – ten years, say (after which rules for price revision apply). The tariffs can vary by type of source. These have worked well for a number of countries, including Germany, Spain, Portugal and Austria, that have rapidly expanded the deployment of renewable sources. They also encourage new firms to enter the market, as they can be sure of the returns to investment. Pricing itself is not the whole story, and access to the grid for new suppliers, including on a small scale and in diverse locations, is important too. So too are the planning and regulatory regimes. Long delays in permissions for nuclear plants or onshore wind can add greatly to costs. All these factors influence innovation and deployment, and thus the extent to which the private sector is able to learn from experience.

Disadvantages of a number of support schemes for the deployment of new technologies, including feed-in tariffs, lie in higher prices for consumers, since it is they who ultimately pay for the electricity. Eventually, however, consumers will benefit from the experience gained in deploying and using the new technologies. Where the experience is of value to a wide range of producers and consumers, it is necessary to consider how broadly across the electricity system the extra costs of supporting these new technologies should be shared. And given that learning processes will benefit everyone by increasing the scale and decreasing the cost of emissions reductions, there are grounds for contributions from public-sector budgets as a whole, not just from the particular industry concerned.

The necessary scale of deployment support for new technologies over the next few decades is likely to be large – potentially much bigger than the requirements for R&D. The IEA has estimated that, in order to achieve emission cuts of the scale required, deployment support over the next twenty years should be around $60 billion a year, double the current sum. Most of this would have to be paid for by consumers. Large though this number is, it is minuscule compared with the roughly $1,000 billion per annum of corresponding investment in energy infrastructure which is, in any case, likely to be necessary in the corresponding industries.[12]

Achieving close-to-zero carbon electricity – the case of Germany

Wind energy has been rapidly growing in Germany for the past twenty years. Germany has gone from having close to zero installed capacity in 1991 to around 23,000 MW in 2007. No country currently has more wind turbines than Germany. Germany's 19,460 wind turbines generated 39.5 TWh of wind energy in 2007. Wind energy alone accounted for around 7% of Germany's total electricity consumption in 2007. German manufacturers have a 37% market share in the global turbine and components market, earning around €6 billion in exports in 2007. This rapid and sustained growth can be attributed to efforts by the German government to support low-carbon electricity generation.

The '100 MW of wind' programme was launched in 1989 and represented a watershed in German energy policy. The Electricity Grid Feed Act of 1991, followed by the Renewable Energy Sources Act (EEG) of 2000, established a payment for renewable energy per kilowatt hour. The EEG has been central to the growth of Germany's renewable energy sector as it provides economic incentives to produce energy from renewable sources. All energy provided by renewables must be accepted by the operators of the electricity networks at a fixed price stipulated in the EEG. A 2008 amendment to the EEG sets the initial remuneration for new onshore wind turbines on land at €9.2 cents per kWh. The tariff will be decreased by 1% every year for new installations. Initial remuneration for offshore wind turbines is set at 15 cents per kWh until 2015. After then it will decrease to 13 cents/kWh for new turbines, decreasing by 5% per year. There will also be remuneration for repowering projects – that is, for replacing first generation wind turbines with newer, more efficient wind technology. For repowering projects the initial remuneration will be increased by 0.5 cents per kWh.

By 2020, the overall installed wind capacity could be 55,000 MW, generating approximately 150 TWh/year and accounting for about 25% of electricity consumption. The cost of constructing wind turbines, however, has risen due to a rise in the price of raw materials (the cost of steel has doubled since 2004, for example). If the 2020 projections are to be reached then policy will have to continue to be designed carefully with tariffs that maintain the attractiveness of wind power. The way forward may be an emphasis on repowering projects that have the potential to double onshore capacity and triple the energy yield with fewer wind turbines. Germany is a clear example of how government policy can support a transition towards close-to-zero-emissions electricity generation.

While electricity supply and transport tend to get special attention in discussions of technological support, it is important to look beyond them to other sectors of the economy. There is huge potential for emissions reductions through more efficient buildings and industry, for example, and recycling is already making a major contribution to keeping down emissions. Indeed, its scale is so little appreciated that it might be described as one of the 'best-kept secrets' in energy and climate change. Recycling saves, globally, emissions equivalent to around two-thirds of the total emissions from air transport; about 40% of raw materials by

Growth of Wind Energy in Germany: Political Milestones

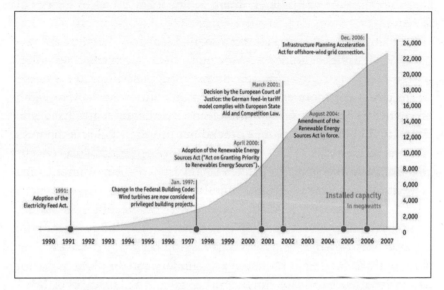

Source: *Wind Energy – an energy source with a fantastic future*. German Wind Energy Association. March 2008.

value come from recycling, and for steel and aluminium the energy saving from a tonne of recycled versus a tonne of new materials is more than 80% and 95% respectively. New technologies for separating out forms of waste could also have a great impact.

Still more market failures

Many other market failures have to do with information and organisational structures. Some, but not all, could be overcome by the market itself, if there were a shift in focus and attention. Others require government intervention.

There are many forms of energy efficiency which can generate great cost savings if only firms had the appropriate information and focus. In their book *Green to Gold*, Daniel Esty and Andrew Winston outline cases where firms have saved billions of dollars from discovering and taking opportunities for energy saving which they had simply overlooked (see Chapter 7). This type of failure to grasp opportunity can be overcome by better leadership, action, management and information. Learning from the example of others will be a key driver of savings in this area.

Problems in capital markets can also inhibit action. They range from the individual householder who cannot obtain a loan for a ground-source heat pump, to consortia of governments who have trouble raising the huge amounts of capital required to move forward research on nuclear fusion. In some cases, these can be solved by financial innovation in the markets themselves, but in others public support is likely to be necessary.

Other problems concern failures to link the creation of assets, such as the improved insulation of an office or house, to the markets for the services of these assets, such as the rental of the office or house. Housebuilders or investors often argue that it is difficult to recoup the costs of the investments involved in making buildings more efficient because those buying or renting do not understand, or perhaps do not have the patience to focus on, the future savings in energy bills. It is a striking failure in the combination of information, perceptions and markets, and/or a peculiar structure of preferences or decision-making which appears to place a high priority on short-term issues and costs.

In these circumstances, simple price mechanisms or corrections do not work well. A more sensible response may, instead, be strong regulations on building efficiency; if the causes are as described, such regulations could make both builders and homeowners better off. As we become more aware of the dangers of climate change, there may be grounds for making these regulations still stronger (see the next chapter for some examples). In other cases, the policies of the 'nudge' may be helpful. For example, letting people know their energy consumption relative to neighbours, or having a red light come on when energy use is high, can 'nudge' people into more careful or responsible action.[13]

Responsibilities, public discussion and preferences

Notions of duty and responsibility underlie crucial aspects of our 'preferences'. Moreover, these preferences may change in response to new information or experiences. If we realise our actions could kill, we might curtail them – drink-driving, for example, has decreased not only because incentives (punishments) were put in place to take account of the externalities (dangers), but also because we understood that it is irresponsible behaviour. Similarly, more people are willing to adjust smoking habits in response to a growing realisation of the possibility of harming others. A similar understanding is developing in relation to emissions from greenhouse gases.

Tobacco, alcohol, other harmful drugs, saving for retirement and physical exercise are all clear examples of issues on which public policy goes beyond the sticks and carrots of price and cost mechanisms, and works, at least in part, via information and discussion. Governments can encourage individuals to behave more responsibly in relation either to their future selves or to those who might bear the costs of their failure to act.

When it comes to climate change, information and public discussion can have a profound effect on what individuals see as responsible. Who takes part in the discussion is also vital, as notions of responsibility depend on which community we believe ourselves to belong to. In this case, the issue concerns the whole world, now and over the indefinite future, so we need to ensure that discussion is broad and that we think through the involvement of future generations. Young people are central to the discussion; we must all try to incorporate their views and think about the rights and preferences of future generations who cannot directly participate in decisions that will affect them.

Communities, urban design and public transport

How towns and cities are structured, and the price of public transport, are clearly vital policy issues with potentially huge effects on the efficiency of the use of time and energy. Congestion of roads in the UK causes estimated annual losses of around £7 8 billion; a lot of these losses could be avoided.[14] This is not only about the construction of the infrastructure for roads, buses and railways; it is also about their pricing and management, the design of cities, the regulations applying to the location of homes, and the use of cars. It concerns the structure of workplaces and practices affecting conventions for physical atten-dance, and of IT and communications facilities affecting whether people have to travel for work. Many or most of these involve networks in some shape or form in which the decisions of an individual on where to live, how to move, how to interact and how to commute have powerful effects on others.

Until now, many aspects of commuting and transport design have been wasteful and inefficient. In the UK, the cost of public transport relative to private transport has risen sharply over the past twenty years, compounding the waste from congestion. Introducing road pricing could decrease congestion by 50% and yield annual benefits of £28 billion in 2025.[15]

Transport and urban design policies must:

1. charge for greenhouse gases
2. price congestion properly so that we make much better use of our infrastructure, while at the same time encouraging shared transport – for example, traffic lanes for cars with multiple occupancy
3. seek out ways of improving energy efficiency much more vigorously
4. invest more strongly, with due care to cost and efficiency, in public transport
5. make walking and cycling easier and safer
6. regulate the expansion of cities and improve design, with still greater focus on the ability of people to use them with as little private transport as possible
7. make it easier for people to generate their own electricity and sell to the grid
8. look at the potential within a community of combined heat and power – local energy structures are important
9. encourage local work practices which allow people to travel less and at more efficient times
10. encourage recycling by making it easier

These ten points all require public or collective action. They have in common organisational and policy requirements that go beyond incentive structures for individuals. As well as simply being more efficient, many or most of them will make our communities more pleasant and safer places to live, and will also foster a more participatory and equitable society. Approaches which primarily involve raising taxes and prices on greenhouse gases can be inegalitarian, as is familiar from all examples of the rationing of scarce resources through prices. In economic policy we usually think of addressing problems of inequality through tax and transfer mechanisms that target income directly rather than through subsidies on particular goods, as the latter are thought to generate more distortions. However, this argument, while valid and important in broad terms, is not wholly convincing. We know, for example, that if we try to address concerns about the unequal distribution of wealth through income taxes and transfers alone, then there will be significant negative effects on work incentives. For example, if we introduce a high tax credit for people on low incomes, this could lead to very high effective marginal tax rates as the tax credit is withdrawn, thus discouraging recipients from increasing the number of hours they work.

That is why it is so important to look for policies that can draw communities together. Public transport is a clear example, but so too are schemes which encourage sharing of private transport, and recycling to allow the reuse of unwanted items, making for a cleaner community. Just as collaboration across nations on climate change can promote action on other crucial issues and make the world a better and safer place to live, teamwork within the community can improve life for us all.

Institutions and coalitions for action

National policies will not work if governments simply introduce them and leave the rest to the market. The actions required to adapt to and limit climate change are so wide-ranging and deep that the whole community must be involved. The same is true internationally, as we shall see later in the book.

Bringing firms and individuals into the construction of policy is crucial. The process of the Grenelle de l'Environnement in France in the middle of 2007 was a striking example of all sections of society coming together to try and find ways forward that are effective, efficient and fair (and *seen* to be fair). The process was launched by incoming President Sarkozy. The discussions included scientists, policy analysts, central and local government, business groups, trades unionists, NGOs and myself. We talked issues through and made suggestions and proposals, some of which are being translated into policy.

It is not only emissions by firms that need to be checked, of course – the performance of nations as a whole must also be guided and monitored. The Climate Change Committee in the UK, for instance, sets targets, measures performance and suggests improvements. In this way the government has established a mechanism to which it is itself accountable. If it rejects the targets recommended by the committee it must explain itself to Parliament. Agreement across all political parties in the UK has been an important element in establishing this mechanism. Government has a statutory obligation to meet the targets it sets. In both rejecting a recommended target and failing to meet targets, the first sanction is political embarrassment. No doubt we will discover through the courts what the further sanctions turn out to be.

The creation and delivery of policy on climate change is in its early stages, and will be sustained and taken forward as a result of a collection of forces and pressures from: citizens, NGOs and firms; leadership from the top; and various political and other coalitions. In the USA we see close working relationships between political structures, relevant firms and environmental NGOs. The Environmental Defense Fund is one of the largest and most effective NGOs in the USA, working closely with businesses and with governments at all levels. Education, and children and university students, will play a central role too – it is their world that will be damaged by climate change and they often have a particularly clear understanding of the issues and challenges. In Europe the EU acts as a regional structure and oversees obligations and commitments. Given that the response must cut across national boundaries, the development of similar structures on other continents is crucial.

There will also, of course, be coalitions *against* action. Those parts of the car industry specialising in large engines and high petrol consumption, for example, may argue that change to much more economical engines will be very costly; those involved in natural resource extraction may see their economic rents fall. Any change will bring some losers, at least in the short run, and policy should be designed to help manage the dislocation of production processes and workers. Change is necessary and dislocation cannot be avoided, but the process should be managed with great care to prevent the burdens from falling disproportionately on vulnerable groups.

All non-hydrocarbon energy sources have their own environmental problems. Many object to the location of wind, solar or nuclear electricity establishments near where they live or work. (When I began work on the Stern Review, the first letters I received were from two aristocrats with estates in central England who 'trusted I would conclude' that a proposed wind farm in their area was unnecessary.) Nuclear power and CCS both have powerful critics within environmental communities who are arguing for strong action on climate change. My own view is that we will need all the technologies available. Each has different pros and cons. We will have to deal with the problems of nuclear waste and take care with nuclear proliferation. Each technology will have to prove itself both commercially and on safety grounds. No doubt different countries and communities will take different decisions; indeed diversity and its associated learning has its advantages. We should not rule any of them out without very careful consideration. And we must also compare the problems of non-hydrocarbon technologies with the huge dangers of continuing with hydrocarbon growth. We need to know now if CCS can work on a commercial scale and whether the transport and storage of CO_2 can be done safely – without CCS for coal and gas, it will be much more costly to achieve the target concentration levels.[16]

All of these issues must be analysed, thought through and acted upon. Policy analysis and research, technological R&D, and action and investment must occur simultaneously and rapidly. There is a wide range of choices and they have to be argued through in each country, with results and experiences shared throughout the world. The design of policies and the challenge of implementation is where economists, other social scientists and policy analysts should now be focusing their efforts.

CHAPTER 7

Individuals, firms, communities: the power of example

Individuals, firms and communities should not just wait for governments to sort things out. There is much that they can and should do to respond to the risks and opportunities that climate change presents, to work together and to mobilise politically to put pressure on governments to act on the scale required.

The focus of this chapter is how to motivate action. Clearly, the right financial incentives must be in place, be they carbon prices, taxes, grants, or other mechanisms – but other, less tangible factors are equally important: access to good-quality information on how to make low-carbon choices; tackling the 'hassle factor' head-on so that it is easier to take up low-carbon options; and, crucially, building a sense of shared responsibility, so that people feel that their own small actions are a worthwhile contribution to a wider collective effort.

The 'power of the example' is an especially important catalyst for motivating action, and is something I have seen throughout my work, from the smallholder tea growers in Kenya in the 1960s, to the wheat fields of India from the 1970s onwards and advising governments as chief economist of the World Bank in the 2000s. The Kenyan women of Kericho, Kisii and Nyeri saw, and learned from, each other growing tea on small plots successfully, a real innovation in the 1960s when previously tea had been almost entirely an estate crop; the villagers of Palanpur in Uttar Pradesh, where I have been working for thirty-five years, learned from each other about new varieties in the Green Revolution of the 1970s; and in response to the description of an option, the finance ministers and prime ministers with whom I discussed policy almost always asked, 'Has someone made this work?' There is no substitute for seeing others do something successfully. The examples in this chapter are just a small selection of individuals, firms and communities who have done just that.

Individual action: identifying opportunities and motivating action

Just as many people do not buy carpets or clothes made from child labour, notions of basic personal responsibility mean that they also wish to cut their emissions to reduce the threats to future generations, including their own children and grandchildren, even though they know that what they can do plays only a tiny role in the process. If they can also save money by reducing electricity bills or the cost of running a car, so much the better.

The annual emissions of greenhouse gases for a typical European lifestyle are, per person, around 10–12 tonnes, the most significant contributors to which come from heating homes, using electrical appliances, and travel by car and air. Across these areas, there are a huge number of opportunities to cut emissions (and save money at the same time). The McKinsey Global Institute found that typical US families who replace incandescent bulbs with compact fluorescent versions could get a payback in less than a year; replacing low-efficiency water heaters with on-demand or solar heaters could save up to 65% of the energy needed, and have an annual return on investment of some 11%, while installing a state-of-the-art heat pump could save 25% on the average annual heating and cooling bill, for a cost of less than $1,000.[1] In new homes, the most efficient heating and cooling package saves 50% of energy consumption required by current standards ($400 annually for an average household).[2] In the UK, the average household can save around 1.5 tonnes of greenhouse gases annually by making their home more energy-efficient; between 2002 and 2005, an estimated 1.1 million households installed or topped up their loft insulation, resulting in an estimated £87 million a year savings in energy costs.[3]

Car travel is a large and fast-growing source of emissions. In the USA, personal vehicle use accounts for around one-fifth of all greenhouse gas emissions from fossil fuels.[4] However, there are many small changes that can be made to increase efficiency. Keeping tyres pumped to appropriate levels can reduce motoring costs by 10%, and using cruise control can add a further 10–15% to efficiency. In developing countries, cars tend to account for a smaller fraction of emissions, but this is changing rapidly as increased wealth makes car ownership more widespread. Technologies such as the hybrid electric car offer significant carbon savings, but the costs are still relatively high, and hybrids remain a tiny portion of the market. However, there are other, often cheaper ways to reduce the environmental impact of driving: using a 1.0-litre rather than a 2.0-litre engine saves about 1 tonne of CO_2.[5] Compared to their petrol equivalents, diesel engines consume up to 30% less fuel and emit about 25% less CO_2. In developing countries, two-stroke engines, which power 'two-wheelers' (motorcycles, scooters) and 'three-wheelers' (tricycles, tuk-tuks), represent one of the largest sources of vehicular emissions.[6] The organisation Envirofit International has developed a Direct In-cylinder fuel injection retrofit kit for two-stroke engines that is cleaner and more fuel-efficient than possible four-stroke alternatives. And simple decisions such as replacing short car trips with walking or cycling can significantly contribute to savings in carbon emissions while at the same time encouraging healthier lifestyles.

Individuals can have an impact by making changes in their food-purchasing habits and diet. A UN Food and Agriculture Organisation (FAO) study calculates that meat production accounts for about 18% of worldwide greenhouse gas emissions compared with transportation, which accounts for 13%. The FAO adopted a broad approach to the consequences of meat production, when arriving at the 18% figure: clearing land, creation and transportation of fertilisers, burning fuels in farm vehicles, and the emissions coming directly from cows and sheep. The 18% is a gross estimate (total emissions), not the difference between an animal-based and a non-animal diet.

A 2005 study from the University of Chicago compared the greenhouse gas emissions of different American diets.[7] It estimates that moving from a 'mean American diet' (which includes 27% meat, by calories) to a vegan diet would save about 1.5 tonnes of CO_2 per person per year. In comparison, switching your Camry for a Prius saves about 1 tonne of CO_2; giving up a diet of 50% red meat and going vegan saves 3.6 tonnes of CO_2 per person per year, similar to switching from an SUV to a Prius; going from 50% red meat to 50% poultry saves some 2 tonnes of CO_2. In his 2008 Climate Change Review for the Australian government, Ross Garnaut drew attention to the environmental and climate advantages of replacing beef or lamb with kangaroo. Meat consumption is significantly lower in developing countries, but the

FAO projects global consumption to rise as population and incomes increase in poorer countries. By 2050, it expects meat demand to be twice the 220 million tonnes the world consumed in 2000.

These are only a very few examples of the actions that individuals can take, often relatively simply and at low cost, but we know that even cost-effective opportunities are frequently not embraced. Incandescent light bulbs, for instance, still make up the majority of the lighting market, despite the fact that there are alternatives which are both more cost-effective and more environmentally sound. Fuel saved by switching to a LED system can add up to a month's earnings each year for people in developing countries who currently rely on kerosene lamps.[8]

Evidence on people's willingness to change their own behaviour is mixed. A survey for HSBC[9] in relation to its Climate Confidence Index 2007 found that more people say they are prepared to make changes to their lifestyle to help reduce climate change (58%), than are prepared to spend extra time (45%) or extra money (28%).[10] The question is, what will it take for individuals to change their habits, behaviour and purchasing decisions?

By making it easy to take action, and by providing examples of how to do it, people can be motivated to make a responsible choice. If they understand the advantages of insulation, if local firms and super-markets make it easy to obtain materials and understand the options, if governments provide simple and rapid processes to obtain relevant grants, if finance is available from banks or other sources, and if people know others who have done it successfully, at reasonable cost and without excessive fuss and hassle, then take-up can move quickly. Not all these factors need to be in place, but each one contributes to the momentum. In addition, action is more likely if individuals, businesses and governments feel they are working as part of a global effort, and do not perceive their actions as 'too small to make a difference' or that their efforts are wasted. The need for collective action is one of the principles underlying the importance of a global agreement, as discussed in the next chapter.

Information on what is possible is becoming more widely available in the UK, through institutions such as the Carbon Trust or books like Chris Goodall's *How to Live a Low-Carbon Life*. Many websites help households work out their carbon footprint, as do some financial institutions and government initiatives like the Act on CO_2 calculator.

Campaigns aimed at consumers, such as the Climate Group's Together Campaign which works with brand-name partners like Tesco and Target, are effectively educating people on easy and affordable ways to take action in their everyday lives.[11] Information is also becoming available that can help individuals reduce their consumption of embedded carbon – that is, the emissions involved in producing goods and services.[12] Cutting down on embedded carbon is difficult for the consumer as it requires knowledge of production processes, but some retailers are currently working to make information more widely available. The chief executive of Tesco, Sir Terry Leahy, has committed to provide comprehensive labelling of the carbon content resulting from the manufacture and distribution of a variety of products. Consumers will also use their own judgement or public information; consumer websites and blogs will act as a check on companies who market themselves as 'green' or low-carbon.

As well as informing people, it is important to make it as easy as possible for them to make responsible choices – simply by providing separate bins for recycling, for example. In many developed countries separating and recycling was rare even ten years ago, but now it is often the norm, and in some places is a legal requirement. There is, though, considerable room for improvement in many countries.

Creating options is of great importance in transport. Encouraging people to use cars less will obviously have little effect in the absence of alternatives, and competitively priced, convenient public transport can transform the way people travel, as witnessed by new metro systems round the world, whether they be in Beijing or Delhi. A better bus service has attracted a big increase in passengers in London. Well-protected cycle lanes and appropriate road rules can make cycling much safer and popular. The publicly supplied Velib bicycle rental service in Paris has been a great success and similar schemes operate in many European cities. In Latin America, Bogotá's CicloRuta is an extensive bicycle-path network which has reduced car dependence and associated emissions.

In addition to lowering their direct carbon footprint, people have the opportunity to offset their emissions. The voluntary carbon market provides a way for individuals to buy greenhouse gas reductions outside the confines of the EU and Kyoto schemes. There have been debates about the credibility of these programmes, and there is certainly a need for more robust standards and institutions to regulate the market, but good providers already exist: Climate Care (www.climatecare.org), for example, is one my family uses.

Again, offsetting needs to be easier if its take-up is to increase. The actions of firms can make a big difference here. Andy Harrison, the CEO of easyJet, said at a seminar on the Stern Review in November 2006 that his company aimed to have 50% of customers offsetting their flights within five years; it is now offered to customers at the point of booking, alongside familiar options such as travel insurance. Similarly, some supermarkets are considering offering a service whereby customers can offset their emissions at the checkout. By going down these routes, firms can help customers feel more engaged with the environmental agenda, as well as bolstering their own reputation as responsible companies.

In developing countries, innovative ways of encouraging individual action are also emerging. In India, The Energy and Resource Institute (TERI), in collaboration with the Ramakrishna Mission, is working with women in the Sunderbans in the southern region of Gangetic West Bengal in a pilot project to promote solar power. The enterprises provide solar PV-based services in remote and interior villages, as well as repair and maintenance services to existing products and systems. With the support system behind them, the women charge and rent solar lanterns on a daily basis, design and assemble small electronic items and repair solar home systems. As a result, they gain financial independence while also promoting innovative change with long-term benefits for their community.[13] In partnership with the Climate Outreach to Youth in India Programme, TERI also promotes awareness of policy issues related to climate change among schoolchildren in New Delhi through 'climate clubs', workshops and guidelines for teachers.

For many individuals and households, the crucial issue of adapting to climate change will involve understanding the increasing threats of droughts, floods and storms. There is a great challenge in providing that information at a local level, as was emphasised in Chapter 4, but if people have that knowledge, there is much that they can do defensively, whether through building design or even choosing where to live. Well-integrated information and communication technologies (ICT) networks reduce inefficiency and waste as well as inform populations of risks and vulnerabilities. For example, networked ICT has the potential to improve the efficiency of electricity generation and use, provide better integrated transport facilities and reduce the need to travel by allowing workers to use more convenient locations. By providing real-time monitoring of water movements, mountain ice melt and land-use productivity, as well as by connecting local communities to government institutions, ICT can enhance adaptive capacities and improve disaster prevention and responsiveness. As insurers recognise the realities of climate change, revisions in insurance coverage and pricing are forcing individuals to react. Other insurers are insisting that policyholders take defensive measures such as installing storm shutters before they are prepared to provide coverage.[14]

Adaptation issues are even more immediate for people living in developing countries. Those living in a poor rural area or an urban slum are far more familiar with the dangers of severe weather patterns than people living comfortably in the rich world. Even though Hurricane Katrina got more attention, the monsoon flooding in Mumbai, India, the previous month killed a similar number of people (around 1,000), and Cyclone Sidr in Bangladesh two years later killed far more (around 3,000) and displaced 7 million people. The last ten years have seen devastating floods in Venezuela, where more than 30,000 died in 1999, in Mozambique in 2000, and in Central Africa and Bihar in India in 2007. Droughts in Ethiopia in 2003 and in northern Kenya in 2006 caused profound hardship and thousands of deaths. The rural population in north India is enormously vulnerable to changes in the summer monsoon which provides their livelihoods in terms of water for the rest of the year, but can also unleash destructive flooding.

Despite these traumatic events, understanding of the links between climate change and physical risks is much less widespread in the developing world, and is a major obstacle to the ability of poor rural households to strengthen their resistance to the threats. The UN has developed several education programmes such as the UNEP Climate Outreach Programme supported by the UNFCCC, a project which advises governments on how to promote awareness at the national level, and supports NGO efforts to provide accurate and accessible messages concerning implications of the work of the IPCC. With a focus on learning from experience, the UN has also developed a website to promote the sharing of information around adaptation: the Adaptation Learning Mechanism (www.adaptationlearning.net), which is a $1 million project capturing and disseminating adaptation experiences and case studies.

Political pressure and public opinion

As well as taking action to reduce their own emissions, individuals, either separately as voters and citizens, or together through civil society including NGOs, can put pressure on governments to establish a climate policy framework and to agree a global deal.

International pressure groups such as Friends of the Earth, WWF and Greenpeace have of course been active on these issues for many years, and as well as traditional campaigning and direct action, many are now striking up alliances with other parts of society in order to move forward. For example, the WWF is working with a number of major global businesses in the Climate Savers programme, which aims to promote efficiency and cut emissions.[15]

Individuals can also have a powerful influence. Nobel prizewinner Al Gore's Oscar-winning film *An Inconvenient Truth* has had a great international impact recently, as have Nicolas Hulot, a prominent and very effective environmentalist in French television, and David Suzuki, a remarkable scientist and commentator in Canada, both within their own countries and more widely. Hulot, for instance, was instrumental in making climate change an issue in the 2007 French presidential election, by asking all the main candidates to sign his 'Pacte écologique' and threatening to run for president if they did not: they all signed. Yann Arthus-Bertrand, a well-known photographer and author of *The Earth from Above*, initiated an environmental awareness campaign for students in 50,000 schools across France.

The media, newspapers, radio, television, Web and so on have a great responsibility in presenting evidence in a measured and careful way. All too often, however, a desire for theatre or misplaced assessment of the balance of the argument leads them to, for example, give similar time to scientists and deniers of the science, when the balance of the argument in logic and evidence is 99 (or more) to 1, not

50–50. There is also a real difficulty in keeping long-term issues in the public eye.

The evidence on public opinion is encouraging. In 2006, the Chicago Council on Global Affairs and worldpublicopinion.org, in cooperation with global organisations, found that in all countries where polling was undertaken, a significant majority see climate as an important or critical threat, including in China and India.[16] People also see the need for a strong response; according to a 2007 BBC World Service poll of 22,000 people in twenty-one countries, nine out of ten said that action is necessary to address global warming and a substantial majority (65%) chose the strongest position which calls for major steps starting very soon.[17] The countries with the largest majorities favouring the toughest action on climate change are in Europe: Spain (91%), Italy (86%) and France (85%). Latin American countries were also strongly in favour: Mexico (83%), Chile (78%) and Brazil (76%). Also in 2007, however, HSBC found relatively low levels of confidence that action would be taken – although levels of optimism were relatively high in India (45%) and China (39%), they were low in most other countries, particularly France (5%) and the UK (6%).[18] There is a growing understanding of the importance of action, but much to do in order to galvanise it.

In line with strengthening public opinion, voters have made themselves felt on climate issues. In California, Governor Schwarzenegger has been a leader in climate change policy for his state, putting in place an ambitious target for emissions reductions of 80% below 1990 by 2050, and has taken a leadership role in the national debate.[19] He was re-elected in 2006 with an increased majority. In November 2007, the voters of Australia comprehensively voted out the government of John Howard, motivated in part by the hostile line he had previously taken on action on climate change and his refusal to sign the Kyoto Protocol. Kevin Rudd replaced him and within weeks had signed Kyoto and played a prominent part in the UNFCCC meeting in Bali in December 2007 which launched the negotiations for the successor to Kyoto. Ultimately, if public opinion on climate change action is strong, then politicians will listen. However, the causality runs both ways: clear and decisive leadership by legislative and business leaders can inform and influence public opinion, so that a more engaged public in turn press for greater action from their representatives.

Firms: opportunities, risks and leadership

Firms are well placed, both to take action and foster it. They can educate clients and employees on the business implications and develop products and policies which enable them to take effective action. And firms are now starting to see climate change not just as an issue of corporate responsibility, but increasingly as something that will affect their bottom line in a tangible way. Regulations on greenhouse gases will increase the costs to energy-intensive businesses, while growing public awareness means that climate change is becoming a branding and reputational issue. The importance of responsible action on climate change for companies is something I see from my own experience as a university teacher: students increasingly steer away from firms they regard as environmentally irresponsible, in much the same way as they might avoid tobacco and arms companies. The way businesses behave affects the quality of the people who want to work for them. The same is true for many potential investors.

Firms also face opportunities, of course, as markets for low-carbon goods and services expand. A 2008 McKinsey Global Survey found that 60% of global executives view climate change as important to consider within their companies' overall strategy.[20] The more progress is made on global action, the more costly it will be for firms to ignore the issue. As firms perceive that the world is becoming more carbon-constrained, the advantages of managing that transition and availing themselves of new opportunities are compelling.

Companies, nonetheless, face a challenge to translate the importance they place on the issue into concrete business action. The same McKinsey survey also found that 44% of global executives say that climate change is not a significant item on their agenda. The survey found that many executives are optimistic, however, about the effects of action, with 61% viewing the issues associated with climate change as having a positive effect on profits if managed well.

Increasingly, companies are announcing their commitment to reduce their carbon footprint voluntarily. In doing so, they are discovering new energy-efficient ways of conducting their business. An example of a joint effort is an initiative whereby a group of thirty-seven of the UK's leading food and consumer goods companies (Boots, Marks & Spencer, Unilever, Waitrose and others) have recently attempted to pool their delivery operations and share vehicles on certain routes.[21] This initiative can lead to significant cost savings for each company involved, while also responding to the growing awareness among consumers of the number of food miles embedded in the distribution of products.[22]

As regulation of greenhouse gases intensifies, so the incentives to improve energy efficiency will increase. An example of a company which is tackling efficiency issues across the business is Wal-Mart. The company's then chief executive, Lee Scott, was so influenced by the impact of Hurricane Katrina that he launched an ambitious package of targets and initiatives in 2005 to make Wal-Mart a global leader on climate change action. Action so far includes a 70% saving on energy in supermarket aisles by putting doors on refrigerated cases. In October 2008, Scott told 1,000 of his suppliers in Beijing that Wal-Mart would be demanding strong standards on the environmental and social aspects of the manufacture and distribution of the products they sell, right through the supply chain. Given that Wal-Mart accounts for around 10% of all American imports from China (around 3% of China's total exports), the effects will be radical.

In the UK, British Telecom has reduced its emissions by 35% from 1996 levels by introducing digital technology and switching its fleet from petrol to diesel, for example. Between 2002 and 2006, the savings from better use of energy were £1.5 billion. DuPont, the major international chemical company based in the USA, has found savings of $2 billion per annum from careful scrutiny and action on waste, particularly concerning energy.[23]

Companies are also starting to differentiate themselves from others through their response to climate change. Marks & Spencer aims to become carbon neutral by 2012 and its 'Plan A' programme is driving innovative actions across its product lines. As part of its efforts to reduce its direct carbon footprint, Marks & Spencer is utilising food waste, to create renewable electricity which is being used to power some of its stores. The plan is now a major part of the company's brand identity and marketing strategy. Conversely, a perceived failure to take climate change seriously can expose a company to boycotts, protests from activists or shareholder rebellions. Greenpeace's long-standing 'Stop Esso' campaign, and recent protests in the UK against utilities operating coal power stations, are just two examples. Britain's largest power station, Drax (sometimes nicknamed 'Drax the Destroyer' after a comic boy character), has repeatedly been the target of campaign groups for its high level of CO_2 emissions. At a more detailed level, NGOs are publicising (through Web links, for example) details of where companies source their inputs and products, and the environmental standards they observe. There are fewer and fewer hiding places for firms wanting to conceal dubious, unsafe or irresponsible practices.

As well as reputational issues, there are physical risks to businesses from climate change, the effects of which will transform the attractiveness of key locations and sectors. The insurance industry could face significant losses from extreme weather – the amount of insured loss has already increased from roughly $5 billion a year in 1970 to about $34 billion a year between 2002 and 2006.[24] A number of insurance companies are responding proactively to the changing risk profile they face; they have to, that is their business. Companies such as reinsurer Swiss Re and trade bodies such as the Association of British Insurers have put significant resources into estimating the impact that climate change will have on the business landscape for insurers; their work will not only help the industry to respond, but will also be invaluable in planning adaptation responses by individuals and governments. Munich Re, as well as carrying out its own extensive analysis on climate change, has been a major backer of research, including at my own university, the London School of Economics. In 2007, following an initiative by the Prince of Wales, a group of global insurers joined together in the ClimateWise initiative, pledging to incorporate climate change into their business strategies.

But as well as risks, there are opportunities. The transformation of energy systems from high- to low-carbon creates new, multibillion-dollar markets, which are now attracting significant capital. Investments in clean energy in 2007 grew 41% over the previous year to around $117 billion.[25] On one estimate, the market for wind power will grow from $30 billion a year now to more than $80 billion by 2017, and for solar PV from $20 billion to more than $70 billion over the same period.[26] Some of the funding for companies in these markets is now being channelled through specialist venture capital or equity funds, attracting investors interested in taking advantage of the growth opportunities. HSBC, for instance, launched a climate change benchmark index in 2007, which identified three hundred companies contributing to action on climate change; from 2004 to the point of launch, these companies outperformed the mainstream global equity index by around 70%.

A committed transition to a low-carbon economy can usher in a new era of innovation and creativity. An example which may become important is biochar, which is made by heating biomass in the absence of oxygen, a process known as pyrolysis. When buried, biochar enhances the productivity of the soil. Thus the absorption of CO_2 takes place when the biomass is grown but carbon dioxide is not released by burning the biomass either directly or as biofuel. And it can substitute for fertiliser made by hydrocarbons. It is better than carbon neutral: it is carbon negative, in the sense that, net, it removes CO_2.

Vinod Khosla, the venture capitalist who is a leader in clean-technology investments, has recently invested in Calera, a company which is developing new building products made from CO_2 captured from the atmosphere or from electricity generation. This is CCS in which the carbon is not stored as a liquid transported by pipelines and placed deep underground, but as a solid which is useful in construction. If successful, the potential is immense.

By the time this book is published, there will be countless additional technological possibilities. The flow of ideas is remarkable. After many speaking engagements, I leave with a pocketful of business cards associated with new ideas. If only a fraction of these ideas represent viable possibilities, the reality of low-carbon growth will be economically secure.

Carbon markets will also play a pivotal role in this process. In 2007–8, carbon trading volumes rose 80% to $60 billion, nearly two-thirds of which was associated with the EU ETS.[27] Global income streams linked to climate change will continue to grow. Capital markets are already anticipating the opportunities, and leading financial institutions – Barclays, Deutsche Bank, Merrill Lynch and UBS – are positioning themselves for areas of rapid growth. Goldman Sachs bought 10% of Climate Exchange plc (CLE) shares in 2006, while in 2007 Morgan Stanley set up the Morgan Stanley Carbon Bank. As emissions trading spreads throughout the world, so the role of financial intermediaries will become ever more important. Carbon is becoming a new asset class for the portfolios of financial institutions.

Low-carbon investments are currently greatest in developed markets, but there is strong growth potential in the developing world. A recent Climate Group study, 'China's Clean Revolution', found that China invested $12 billion in renewables during 2007, second only to Germany.[28] China is now home to a number of world-leading renewable energy companies. Suntech Power Holdings, for instance, has 5% of the global market share in solar power, and is at the forefront of the development of building-integrated solar technologies, which build solar panels directly into materials such as roof tiles.

With climate change having an obvious impact on company performance across sectors and across countries, shareholders are increasingly demanding that firms provide information on how climate change will affect them, and that they factor it into their strategic planning. Often, even basic information on emissions is lacking. The Carbon Disclosure Project, an NGO which acts as an intermediary between shareholders and corporations on climate-change related issues, works on carbon disclosure methodology and processes, seeking information on the business risks and opportunities presented by climate change and greenhouse gas emissions data from the world's largest companies (3,000 in 2008).[29] Sometimes shareholders will take more direct action – in 2008, a record fifty-four US companies faced shareholder resolutions calling on them to respond to climate change risks and opportunities.[30]

Climate change is no longer an issue that companies can choose to avoid, whether in 'good times' or in an economic downturn. Forward-looking companies who have demonstrated a commitment to the environment and are viewed as acting responsibly are better placed to retain the trust of consumers, shareholders and government, as well as to differentiate themselves from competitors. In a downturn, companies may also see new business developments in clean energy as increasingly attractive new opportunities. As Yvo de Boer, head of the UNFCCC, suggested, 'The credit crisis can be used to make progress in a new direction, an opportunity for global economic green growth.'[31]

The role of firms in policy

Firms can do much to respond to climate change through their own actions, but without a strong public policy framework in place, the incentives to act will fail to match the scale required. There is much that businesses can do to encourage governments to put such frameworks in place, in a way that is cost-effective. Companies have the ear of national governments, and are also well placed to understand what kinds of regulations and policies will provide incentives that will genuinely address market failures and encourage innovation. Often acting together, companies are increasingly playing a positive role in the policy dialogue. At the beginning of 2007, a group of business leaders from some of the biggest firms in the USA, including Alcoa Aluminum, BP America, Caterpillar, DuPont, General Electric, Lehman Brothers and three of the country's largest utilities (Duke, PG&E and FP&L), formed the US Climate Action Partnership and wrote to President Bush asking for clear policy signals on climate change. Their intervention set the stage for a softening in US policy on climate change, with Bush unambiguously accepting the need for action in a speech shortly before the 2007 G8 Summit.

Another example was the role of UK business in shaping the country's allocations plan in the EU ETS. In Phase 1 of the scheme, the UK government, like all other EU governments, was extremely cautious in its allocation plans in the face of strong business lobbying, in order to avoid accusations of putting jobs at risk. In the second phase, however, a group of major UK companies lobbied in favour of a more ambitious approach, arguing that a stable and strong carbon price was essential in order to give business sufficient certainty to make long-term investment decisions.[32] In response, the UK government implemented a strict plan, contributing to the success of Phase 2 of the scheme in achieving a stronger carbon price.

Also in the UK, the Confederation of British Industry (CBI) published an excellent analysis of technological and policy options in September 2007. The report, by eighteen chairmen and chief executives of firms including British Airways, Tesco, BT, Shell and Ford, made a clear call for serious and immediate action on climate change. In June 2008, and in spite of the economic slowdown, the CBI further called on business to continue the drive to improve its environmental performance and reiterated the importance of a carbon price.

Similarly, in developing countries there has been an increased awareness among businesses. In February 2008, the Confederation of Indian Industry published a valuable document, 'Building a Low-Carbon Indian Economy'. The paper lists actions and a large number of technological measures to be adopted in the short, medium and long term in various industries – aluminium, cement, ceramics, glass, pulp and paper, co-generation steam and condensate systems, sugar, textile, foundry, iron and steel, fertiliser and engineering – to achieve higher energy efficiency.[33]

At the international level, in 2007 the chief executives of 153 companies worldwide, including thirty from the Fortune Global 500, committed to speeding up action on climate change and called on governments to agree as soon as possible on measures to secure workable and inclusive climate market mechanisms post 2012. The Business Leaders Call for Climate Action was made in a statement issued ahead of the Bali UNFCCC meeting.[34]

Other international initiatives focus on particular issues, such as the Prince's Rainforest Project. Set up in October 2007 by the Prince of Wales, the project undertakes a number of initiatives seeking to raise awareness of the need to protect the world's remaining forests and to identify and promote practical methods to halt deforestation.[35] The project aims to work alongside other governmental and non-governmental initiatives, including the UNFCCC. A number of firms are involved, including McDonald's, Barclays and Shell.

There are times when interventions are less clear and positive. Some US companies, in particular ExxonMobil, were very active during much of the Bush administration in lobbying to slow down or avoid action on climate change, either directly or by funding climate-sceptic think tanks. However, with even the US government accepting the science of climate change and the need for action, fewer and fewer companies now attempt to deny it; in 2007, ExxonMobil finally dropped its support of two controversial think tanks. Other companies play, usually without much credible analysis, on the problems of competitiveness and potential job losses. For instance, German manufacturers such as BMW and Mercedes tend to produce cars with larger engines and relatively higher emissions, and perceive potential losses from greenhouse gas legislation. Highly sensitive to this fact, the German government is pushing the EU to delay and weaken proposed legislation to limit car emissions. One of Germany's conditions is that all categories of cars should have to cut their emissions – including smaller, lower-emission vehicles produced by France and Italy that already meet the EU criteria on emissions per kilometre – and that the mandatory system should be phased in more slowly than planned. But here, too, company positions are evolving – BMW, for example, has invested heavily in cleaner technologies and hopes to make this a source of competitive advantage in the future.

In many ways, industry has been ahead of government in looking to the long term for the analysis of policy options, risks and opportunities. While some governments have short-time horizons – primarily the next election – firms making long-term investment decisions have to think of horizons over a few decades, and thus have every reason to think through where policy is going and to contribute to its sensible formation.

Communities

Just as individuals and firms, acting alone or collectively, can both reduce their own emissions and put pressure on governments to act, or participate effectively in policy formation, so too can communities, working as towns, cities, states, or in rural areas. Communities can set their own targets and show their feasibility, and thus provide examples which can be a powerful influence on national government policy.

I have experienced this in West Sussex in the UK, where I live. At community meetings, topics range from ground-source heat pumps to cycling, local produce and diet; local business people and NGOs often join the debate. The existence of such groups means that local politicians recognise the issues and the interest in climate change as a political reality.

There are many examples of influential leadership in towns and cities. By decentralising energy sources, the small town of Woking in the UK has been able to create energy where it is used, for example from local windmills, thereby reducing transmission losses and increasing efficiency. These savings make small-scale power-generation technologies competitive with grid electricity. Woking also has the first council-owned fuel cell, which produces electricity from natural gas through a chemical process, to power a leisure centre, pool and park, and has promoted the use of heat generated as a by-product of electricity generation to heat local firms and households – another example of potential gains from decentralised generation. Emissions in the town have been cut by 82% from 1990 levels.[36]

'One of the world's largest, oldest and most successful examples of combining heat and power is the Copenhagen district heating system, which supplies 97% of the city with clean, reliable and affordable heating. Set up by five mayors in 1984, the process simply captures waste heat from electricity production – normally released into the sea – and channels it back through pipes into homes.'[37] The system cuts household bills by €1,400 annually, and has saved the Copenhagen district the equivalent of 203,000 tonnes of oil every year – that is 665,000 tonnes of CO_2.

There are also some very creative examples of plans for completely new cities which are emerging. Dongtan in China aims to be the world's first purpose-built eco-city. Its goal is to be as close to carbon neutral as possible, with city vehicles that produce no carbon or particulate emissions and highly efficient water and energy systems. And in the Middle East, the planned city of Masdar, designed by the British architectural firm Foster + Partners, is conceived to rely entirely on solar energy and other renewable sources, with a sustainable, zero-carbon, zero-waste ecology. The Abu Dhabi Future Energy Company, which is behind the project, aims to position Abu Dhabi, a leading Middle Eastern oil-producing nation, as a research and development hub for new energy technologies.

City leaders are also proving proactive in grouping together to share ideas and experience. In 2007, a project was launched bringing together a group of forty cities (the 'C40' cities) committed to taking action to cut emissions. The initiative works in partnership with the Clinton Foundation. By collectively setting standards and seeking out the relevant technologies and equipment, cities can promote the development of new products, techniques and facilities, and lead by encouraging economies of scale and creating big markets, helping to bring costs down. The C40 cities, for instance, are now exploring the possibilities for joint public procurement.

On a larger scale, the idea of low-carbon zones in China, currently under discussion, could become a powerful example for the rest of the world. This type of regional experimentation is part of China's history: the local experiments in the late 1970s and early 1980s, in moving to the household-responsibility system in agriculture from the commune, were hugely successful and quickly went nationwide, transforming China's agriculture with dramatic increases in productivity. Local low-carbon zones in China could be sufficiently large to work with other parts of the world on carbon trading and the sharing of technology.

In the US, with national policy limited by the Bush administration, action at state level has been critical in leading the country's response to climate change. Ten eastern states have joined the Regional Greenhouse Gas Initiative (RGGI), which developed a model law to implement a carbon cap-and-trade scheme covering utilities. The outcome of the RGGI's first cap-and-trade auction in September 2008 was promising, with all 12.5 million carbon permits being sold. In June 2008 California took major steps with an ambitious plan for clean cars, renewable energy and stringent caps on big polluting industries. The plan is the most comprehensive yet by any US state. In addition, the leadership of California has encouraged action across the border: six Mexican states – Sonora, Baja California, Chihuahua, Coahuila, Nuevo Leon and Tamaulipas – have joined together with PG&E (California's main electric utility) and the California Climate Action Registry and signed an agreement to cooperate on efforts to cut emissions.

PG&E has also signed a major deal to develop a pair of solar power plants producing power equivalent to that from a single large coal-fired plant or a small nuclear generator – according to the Environmental News Service, it will generate twelve times more power than the next largest solar plant in operation. More generally there are regulatory incentive structures which encourage PG&E to help improve energy efficiency and thus sell less energy, rather than the conventional structures where there typically is an incentive to sell more.

While some states are trying to take a leadership role, there are constitutional limits on their power and federal laws can pre-empt state laws. However, in a groundbreaking case in 2007, *Massachusetts v. Environmental Protection Agency*, in which twelve states and several cities brought a suit against the EPA, the Supreme Court decided 5–4 to force the EPA to regulate CO_2 and other greenhouse gases as pollutants.[38]

This chapter shows what can be achieved, both practically and politically. It is impressive how many fine examples are available given that public consciousness of the perils of climate change has evolved only relatively recently. There are fundamental lessons here: first, political pressure, leadership and local collaboration can translate into action; second, there is vast entrepreneurship and creativity which can be released by a policy framework which provides the right incentives and priorities; third, the 'power of the example' is of real importance. The potential for effective action is immense, and individuals, firms and communities are at centre stage.

CHAPTER 8

The structure of a global deal

An effective response to the global challenge of climate change requires international collaboration on a scale which is unprecedented. It requires a global deal. Most of the key elements of a 'global deal' have been assembled in the previous chapters. In the first part of the book we examined risks, costs of actions and technologies to arrive at global targets and action plans over time. In the second part of the book we have so far focused mostly on national policies and action by individuals, firms or communities. The task now is to put the components together in a way that gives structure to an international agreement, and add necessary detail on a number of these elements. This will, I hope, provide clarity and specificity on where we, as a world, need to go and how we should act together. In other words, this chapter provides the international part of the 'blueprint' that the title of the book promises.

This is not, however, an attempt to provide a full treaty document or to suggest there is only one possible structure for an agreement. What matters at this stage is a clear understanding of the principles underpinning a deal and their basic implications for policies and action. The fine detail of the agreements – in whatever form they emerge, treaty or otherwise – will have to be developed intensely, closely and collaboratively. But all too often such detailed negotiations take place without stepping back and first settling principles, building common understanding and providing a sense of direction. It will be government representatives who negotiate the final detail: our purpose here is to provide a principled and analytical platform for those negotiations. Late-night compromises on the finer points will be much more balanced and reasonable if the sense of direction is clear, if all parties see the gains to cooperation and if the consequences of actions are understood. As set out in chapter six the overarching guiding principles are effectiveness, efficiency and equity.

The building of a shared understanding of these principles and their implications is extremely urgent. The deadline for the negotiations launched at the UNFCCC conference in Bali in December 2007 is the meeting in Copenhagen in December 2009. Copenhagen will be an event of great significance in the process of negotiation and should produce detailed language for an international agreement on emissions-reduction targets and the supporting mechanisms and institutions to achieve those targets. If the specific words that form the substance of agreement are to embody these basic principles and their implications in a structured, coherent and productive way, then the platform of shared understanding must be built by the summer of 2009. As the world struggles with financial and economic crises it is important that this agreement is not delayed.

We return to the challenges of building and managing the details – while securing political support for this process within a very compressed timescale – in the next chapter. There will be some who think that a global deal is unnecessary and undesirable because climate change is at most a minor issue. The confusion and errors in their arguments have been dealt with earlier in the book. There will be others who think that if the USA leads the way, others will follow because they will 'see the light' or be undercut by wonderful new technologies. It would be nice to think that might work, but it seems highly implausible to me. First, political agreement on strong action is unlikely within the USA unless it can be shown that others are likely to follow. Second, it would be *inefficient* to try to make all the running in one place. Further, the pace of change both in the USA and in the following by others would be likely to be far too slow in the absence of explicit under-standings. There seems little alternative to some kind of understanding between nations if they are to be able to move forward with the pace required. This need not, however, require heavy international sanctions, as we shall see in the next chapter. Although international collaboration might seem potentially messy, it does not have to be so, and without it I do not think we shall see the necessary action.

The key elements of a global deal[1]

The deal should be considered as an integrated package, which is necessary if the essential principles of effectiveness, efficiency and equity are to be met. This is not a menu from which selections can be made or from which 'nuggets' can be extracted. I emphasise this strongly for two reasons. First, because the logic requires the deal to be seen as a whole with the different elements supporting each other. Second, because during my time in public life I have often seen politicians or senior civil servants from a number of different countries view a package as a list of possibilities or a chocolate box from which choices might be made. If negotiations take the form 'I will drop this if you drop that', the deal would be highly unlikely to deliver on the scale required, and would probably also be inefficient and inequitable.

The structure of a global deal – the key elements

Targets and trade

- At least 50% cuts in world emissions by 2050 relative to 1990; developed countries to lead in setting country-level targets now.
  - Agreement by developed countries to take on immediate and binding targets of 20% to 40% by 2020, and to commit to a reduction of at least 80% by 2050.
  - Within this time frame developed countries clearly and convincingly demonstrate that low-carbon growth is possible and affordable, including sharing technologies and creating trading and other financing mechanisms.
- Developing countries commit, subject to performance by developed countries, to taking on targets at the latest by 2020.
  - As soon as possible, developing countries put forward credible plans to reach 2 tonnes per capita per annum by 2050, factoring in a peak in emissions before 2030; fast-growing middle-income countries to peak around or before 2020.
- Country emissions reductions and carbon trading schemes to be adopted, which are designed to integrate trading mechanisms with other countries, including with developing countries both before and after their adoption of targets.
  - The supply of carbon credits from developing countries simplified to allow much bigger markets for emissions reductions: with good design, 'carbon flows' likely to rise to $50–100 billion per annum by 2030.

Funding

- Strong initiatives, with public funding, on capacity to halt deforestation and integrate into development programmes, plus preparation for including avoided deforestation in trading. Policies and strategy to be determined by countries where the trees stand. For $15 billion per annum, the world could have a programme which might halve deforestation, as a near-term milestone on the way to halting it. Importance of global action and involvement of international financial institutions in pilot projects, in investment and in coordination of action.

- Development, demonstration and sharing of technologies: (i) diffusion of existing technologies, e.g. wind power; (ii) developing and scaling up near-commercial technologies, e.g. $5 billion per annum commitment to feed-in tariffs for CCS for coal could lead to thirty-plus new commercial-size plants in the next seven to eight years; (iii) creating breakthrough technologies, possibly to include advanced solar, enhanced photosynthesis, algae, carbon storage as usable solids, nuclear fusion.

- Rich countries to deliver on Monterrey (2002 UN), EU 2005 and Gleneagles (2005 G8) commitments on overseas development assistance in context of extra costs of development arising from climate change: potential extra costs of development above $75 billion per annum by 2015; climate change to be central in assessment of international development goals beyond 2015, and their funding.

Emissions targets for the world as a whole, for both rich and developing countries, make up the first two elements of the package. The global target is the starting point, for the reasons argued in the first part of the book concerning the overall risk of inaction and the cost of action. Before we turn to the targets for different countries, their rationale and how they might operate, it should already be obvious that some kind of explicit breakdown of the global target is necessary if responsibilities are to be clear. This enables the deal to be effective and allows equity to play a serious role, and further provides a firm basis for trading (and thus efficiency) through enabling the market to function. At the heart of the breakdown should be a commitment from developed countries to a firm target. This is a key condition for developing countries to take on targets at a point in the future, consistent with both their development objectives and meeting the overall global aim.

The third element, the building of an effective trading regime, is crucial both for efficiency and to generate the kind of flows that can provide sufficient incentives for developing countries to commit to an international agreement, without whose involvement any plan will be doomed. Trading is likely to be required for them to take on targets, because they could then foresee financial flows from the combination of trading schemes and worldwide targets that could support their actions to bring down emissions and finance their investments in low-carbon technologies. Strong targets for the rich countries will be necessary to provide the demand for the flows in these markets, and thus to yield a price of greenhouse gases which gives strong incentives for emission reductions. Trading makes lower costs of meeting targets possible by buying reductions where they can be made at lowest cost, and thus could help persuade rich countries to take on strong targets. So the first three elements – global targets, country-specific targets and trading – are intimately linked, both to each other and to the three fundamental principles of effectiveness, efficiency and equity.

The emphasis on trading underlines what should already be clear – a target can be met partly by reduction at home, and partly by buying from abroad. If a target for a country is interpreted as requiring fulfilment solely through action in that country, then the cost of action would generally be much higher, given that cheaper abatement opportunities could arise elsewhere. In other words, a rigid interpretation that action must be entirely domestic will be inefficient. If, at any given level of reductions for countries A and B, the cost of an extra unit of reduction is lower in A than in B, then the total reductions can be achieved at lower cost by doing a little more reduction in A and a little less in B. In these circumstances, B could pay A to do a little more, and both countries would be better off. Alternatively there could be stronger reductions for the same overall cost. It makes no sense to see trading as a cop-out or reneging on domestic responsibility, although one occasionally hears that view expressed.

The fourth element, combating deforestation, is essential. It will be extremely difficult to achieve an overall 50%-plus cut in emissions by 2050 without stopping deforestation (which contributes about 20% of global emissions). The halting of deforestation will also bring great benefits for biodiversity, water, environment and development more generally.

The fifth element, technological advance, is also crucial since without the dissemination of existing technologies and the development of new ones, costs of action will approach levels likely to impede a willingness to act. If technologies are not shared, we will be withholding lower-cost actions from others, thus undermining efficiency. Willingness-to-trade, element three of the package, depends on confidence in access to appropriate technologies. And sharing of technologies on reasonable terms is essential to equity. In other words, without technological advance and sharing, we substantially weaken the whole package and its principles of effectiveness, efficiency and equity.

The sixth element, delivery on overseas assistance for supporting development goals in a more hostile climate, is a basic requirement for equity. The rich countries are responsible for the majority of the increase of emissions in the past, and thus for our difficult starting point. They set a development model for the intensive use of hydrocarbons from which the whole world must now depart. And it is developing countries that are hit earliest and hardest by the effects of climate change. Without strong support for this key area, the perceived injustice will jeopardise the willingness of developing countries to take on targets.

All the elements of the package are mutually reinforcing and arise directly from the three basic principles. Taken together they would demonstrate a mutual commitment to work in a principled and practical way to provide a powerful international response to the two greatest challenges of our age, climate change and the battle against world poverty. This practical manifestation of an international collaborative spirit would itself reinforce the types of attitudes and behaviour likely to sustain any agreement.

1. Global and rich country targets

GLOBAL TARGETS

The first goal is to hold concentrations at or below 500 ppm CO_2e, which means cutting global emissions by at least 50% by 2050, relative to 1990 levels. This target shapes the whole story, the implications of which are at the heart of the negotiations. However, the question of the appropriate target for concentrations has been hotly contested and it is important here to examine the divergence of views.

There are several US economists, a central figure being Bill Nordhaus, who argue for a slow-ramp approach. That is, we gradually increase the price of carbon as we explore technology options and examine damages from climate change more closely. Broadly speaking, versions of this slow-ramp approach, according to Nordhaus, might result in concentrations rising to above 700 ppm CO_2e. I have already explained why such an approach is likely to lead us into extremely dangerous territory from which it would be very difficult to extricate ourselves.

On the other hand, there are a number of scientists, the most prominent being Jim Hansen, who have raised strong and serious arguments to suggest that the target should be no larger than 350 ppm CO_2 (or around 400 ppm CO_2e) given that would bring concentrations back much closer to those in which humankind evolved. The evolutionary processes and the ways in which the physical and human geography have developed give rise to living and settlement patterns for humans and other species which are governed by a particular climate. They point also to the possibility of tipping points such as the collapse of ice sheets, the dying of the Amazon forest, or the release of methane from the permafrost, which could lead to an accelerated process of climate change.

These scientific arguments are very powerful in terms of bringing home the magnitudes of the risks. They convince me that 500 ppm CO_2e, with its high probability of exceeding 2°C above pre-industrial times (96%) and a 44% probability of being above 3°C would indeed be a risky place to be. The problem with a target of 400 ppm CO_2e is that we are already at 430 ppm CO_2e (around 380 ppm CO_2) and we are adding about 2.5 ppm per annum. We are unlikely to turn these additions into negative numbers for a very long time. We will surely be at 450 ppm CO_2e within ten years.

To push hard for a lower target could disrupt the possibility of agreement in the very near future. It might look to some people like an abandonment or reversal of growth and development, and we risk appearing to ask for the impossible. Or we give the impression that a higher target is too unambitious to be worth the trouble since it would, in any case, leave us 'doomed'. On the other hand, a higher target may well look extremely unambitious or even reckless from the point of view of the science. How do we go forward in overcoming these problems and determining targets?

Let us be clear that a target of holding below 500 ppm CO_2e is very strong in terms of the action necessary to achieve it. We will have to act clearly and radically and learn very quickly to change the ways we source and use energy, and we must halt deforestation. As we do so we will discover new technologies and construct new policies, and communities will find new ways to organise themselves. We can then go on from there to revise our targets to those lower than 500 ppm CO_2e, perhaps sooner than we think.

There is a choice of emissions paths to achieve a given concentration level in the future: one can reduce emissions a bit earlier or a bit later, depending on costs and opportunities, so we must also think carefully about targets before and after 2050. The targets before 2050 are of special importance since this long-time horizon cannot be an excuse for delayed action today. Unless global emissions peak around 2020, and strong declines are achieved globally by 2030, it would not be possible to cut by 50% by 2050.

If we are to see no more than an increase of 70 ppm CO_2e in concentrations (the difference between the upper limit of 500 and our current starting point of around 430), then there are limits on total future emissions over the next century.[2] However, given that holding concentrations below 500 ppm CO_2e requires cuts of around 50% by 2050, most sensible paths that achieve this would be consistent with these limits for total emissions.

Looking beyond 2050, the absolute level of global emissions would have to continue on a downward trend if emissions are to stabilise at or below 500 ppm CO_2e. In the absence of breakthroughs on methods for directly extracting greenhouse gases on a large scale, they would probably have to be halved by 2100 from 2050 levels to stabilise at 500 ppm CO_2e, and still greater reductions would be needed if there is to be a real chance of eventual stabilisation levels lower than 500 ppm CO_2e.[3]

As well as explaining the overall targets for the percentage flow reduction required by 2050, we must also be clear on the absolute levels and the per capita levels, both for the specifics of policy and to understand breakdowns between developing and developed countries.

Global emissions for both 1990 and 2000 were around 40 Gt CO_2e per annum (in that decade, emissions from the former communist countries of Europe and central Asia decreased sharply), but are now over 50 Gt CO_2e. Halving them relative to 1990 implies global emissions around 20 Gt CO_2e per annum by 2050.

The global population in 1990 was around 5.3 billion, and in 2000 around 6 billion; it is now (2008) around 6.7 billion. The world per capita per annum emissions in 1990 to 2000 were thus 7–7.5 tonnes, and are now close to 8 tonnes. By mid-century the world population will probably be around 9 billion. The target for emissions in 2050 given that population projection is, as a matter of simple arithmetic, equivalent to around 2 tonnes per capita per annum. These numbers constitute the basic framework for the targets, both for different countries and for groups of countries. The fall from 7–8 tonnes to 2 tonnes is clearly a major challenge.

TARGETS FOR RICH COUNTRIES

Although the current average is 7–8 tonnes per capita per annum, the differentials across the world are huge. In the USA, Canada and Australia they are over 20 tonnes; in Europe and Japan they are 10–12 tonnes; in China over 5 tonnes; in India under 2 tonnes, and in most of sub-Saharan Africa well under 1 tonne.

If the world average is to be around 2 tonnes per capita, then the emissions from most of the major countries will have to be fairly close to that per capita level. The reason is clear and follows from the basic sums: if there are, say, a group of 1 billion people, out of the likely population of 9 billion, at 4 tonnes per capita, then for a world average of 2 tonnes per capita, there will have to be another group of 1 billion people at zero, or another group of 2 billion people at 1 tonne. The average is the average, and if we have one group above there must be a corresponding group below. Given that going far below 2 tonnes per capita in actual emissions will be very difficult, it implies that we should expect all major countries to be clustered around 2 tonnes per capita if the average is to be achieved.

Some may be a little above 2 tonnes per capita in actual emissions, and others a little below. Trade means that emissions *funded* and emissions *achieved* domestically need not be exactly the same, but each major country, or group of countries, will have to have actual emissions close to 2 tonnes – we are not saying, for example, that a big rich country should have a quota target of 2 tonnes per capita so that it could have *actual* emissions at 5 tonnes and try to buy 3 tonnes; it would not be able to find another country to sell that quantity. But there is another point here of great importance. There is *absolutely nothing* in the argument to imply that the quota target or allowance for each country *should be* 2 tonnes per capita. Indeed, there are powerful equity arguments, given the history, that for rich countries the allowance should be zero (which would mean that it would have to pay for all the actual greenhouse gases emitted), or even negative. For both these reasons – trading and equity – the argument that the world should pursue an outright egalitarian objective to have per capita emissions exactly equal by 2050 is not persuasive and neither is the assertion that all entitlements or quotas should be equal by 2050.

In a basic sense, the atmosphere has limited capacity, rather like a reservoir, if we are to keep risks at acceptable levels. The 500 ppm CO_2e limit on concentrations, and the 430 ppm starting point, means that total emissions over the next century or so are limited. The further 70 ppm limit on extra concentrations indicates, roughly speaking, what is left in the reservoir. The problem is that until now there has been no control over access to the reservoir's finite resources – people and nations have been drinking from it at will. Suppose we date the beginning of the drinking to 1850, when the Industrial Revolution started to move quickly and on a greater scale, emissions began to exceed the absorptive capacity of the planet and concentrations started to rise. To suggest that we should all be entitled to emit roughly equal amounts by 2050 is to say that, at the end of the drinking spree, we should be using glasses of the same size. It is difficult to see this as a particularly equitable division of the entitlements to the reservoir, since this type of equality takes no account of all the 'drinking' that has gone on over the previous two hundred years.

We can say that such a plan for allocating the 'drinking' rights would be more equitable than a process that continued as before, but that is a very weak notion of equity. The argument concerning a 'stock' rather than a 'flow' approach to equity applies whichever start date we use. We might, for example, choose a start date of twenty years ago, when the problem was recognised sufficiently clearly for the IPCC to be founded; or 1992 and the creation of the UNFCCC; or the adoption of the Kyoto Protocol in 1997; or even the Copenhagen conference in December 2009. The notional amount in the reservoir would be

different, but the point would still be relevant. Therefore the suggestion here that *actual* per capita emissions should be roughly equal
across nations by 2050 is *not* derived from an ethical principle of
equality of emissions *rights*. If emissions damage others, potentially
killing some of them, do we have an equal right to do so? That does not
seem obvious or persuasive to me. The argument is pragmatic and
arithmetic: if we are to achieve an average of around 2 tonnes per
capita, and if it will be difficult to be far below, then we cannot have
any major groups far above.

The volume and flows of trading over time would be determined,
country by country, by the difference between the time path of actual
emissions and the time path of allocations of rights to emit. At any
point in time the difference between the two is the amount of trade in
permits and that quantity multiplied by the price of an emissions
permit gives the associated flow of funds in or out of the country. If a
pattern of roughly equal rights to emit was negotiated for the period
around 2050, then it is likely that by that time net trading between
nations would be fairly small. That would not, of course, preclude
sizeable actual trading across borders but it would be two-way.

If the allocation of rights to emit in any given year took greater
account both of history and of equity in stocks rather than flows, then
rich countries would have rights to emit which were lower than 2
tonnes per capita (possibly even negative). The negotiations of such
rights involve substantial financial allocations: at $40 per tonne CO_2e a
total world allocation of rights of, say, 30 Gt (roughly the required
flows in 2030) would be worth $1.2 trillion per annum. It is hard to
predict the outcome concerning negotiations over the allocation of
rights, but in my view it would be very difficult, from the perspective
of equity and history, to justify rights in rich countries in 2050 which
are above 2 tonnes per capita and there are strong arguments for
emissions rights lower than that level.

The 80% cuts, at least, by rich nations by 2050 in both actual
emissions and in emissions quotas is central in the framework for a
global deal. For Europe to get down to 2 tonnes per capita, it would
have to emit only a fifth of current levels. Hence, the target is 'at least'
for two reasons. First, emissions in some nations (including the USA,
Canada and Australia) are way above even the European level. Second,
a stock and historical approach to equity and allocations of rights to

emit implies rich countries taking greater responsibility for cuts than represented by rights roughly equal to 2 tonnes per capita – while actual emissions might be in the region of 2 tonnes per capita, rich countries should still be buying the rights to emit, because their *allocation* would be lower. Recognition of this type of argument was seen at the UNFCCC conference in Bali in December 2007 when Norway and New Zealand pledged to be at zero emissions by 2050.

The allocation of emission rights involves, as we have noted, substantial sums and any scheme will have implications for the world distribution of income, or at least increments to it. The notions of equity that should guide these allocations will be value judgements. In my view, we should look at overall valuations of the world distribution of income and changes to it, rather than taking a narrow view of equity purely in terms of energy or carbon. Notions of inequality and poverty are usually examined in terms of income or wealth on the one hand, or rights to participate in a society, for example in terms of education or voting, on the other. There are a number of relevant dimensions and I see little justification for singling out a particular good as opposed to income and wealth as a whole, except for cases such as empowerment and participation in the sense of education or voting.

Thus it seems strange to me to assert a right to a quantity of carbon as a basic notion of equity, not only because a connection between carbon and energy can be broken, but also because the right to energy itself is not necessarily obvious as a basic right distinct from the purchasing power required to buy food, shelter, energy and other goods more generally. If such a right as a basis for equity for a particular good is asserted, this should be justified by careful argument. I would suggest that any notions of equality and justice in the allocation of emission rights should be embedded in a broad view of income distribution, responsibilities for supporting economic development, responsibilities for past emissions and the damages they have done, and the different kind of instruments for influencing world income that are available. Such instruments would include world trade, sharing of knowledge, overseas development assistance, world health and immunisation, and so on. In this context, a notion of zero or negative emissions rights for rich countries for an extended period of time could have some justification as a means for taking account of past emissions, but this would depend on the context and magnitude of action using other instruments. In thinking about the notion of zero rights, we should recognise that equal rights of access to the commons, here the atmosphere, is not the same as equal rights to damage the commons, here polluting the atmosphere. The assertion of equal rights to emit carbon makes four distinct, although partially overlapping, confusions: between energy and carbon; between stocks and flows; between different bases for assessing equality; between rights of access and rights to damage.

A target for 2050 anchors interim targets, and it is vital that substantial reductions are made en route. It would be very costly to try to achieve most of the cuts in the last ten years of the period, as cost is linked to the pace of change. At the same time it is likely that more cuts can be achieved in the second half of the period than the first, because investment with longer lifetimes comes to fruition, there is time to phase out older, more polluting plants, and new technologies will become available. We must also take account of the fact that it is the total emissions between now and 2050 which will determine how much concentrations rise, which indicates the dangers of leaving too much until the last years.[4]

The year 2020 is halfway between 1990 and 2050, and ten years from the December 2009 Copenhagen conference. Strong interim targets of around 20–40% cuts by 2020 for rich countries should be a clear objective en route to their 80% cuts by 2050. Given the increases in some countries since 1990, trading is likely to be an important element in reaching such 2020 targets. In some cases this 2020 target might be slightly lower if 2030 targets are strong – that, no doubt, will be part of the negotiations. Without such types of commitment, given the ease of making statements about targets for far-off dates, it will be very difficult for developing countries to regard the rich world's leadership and intention to act as credible.

2. The role of developing countries

TARGETS, CONDITIONS AND TIMING

The people in developing countries represent the big majority of human beings: 5.7 billion in a total world population of around 6.7 billion now, and will be 8 out of 9 billion in 2050. Emissions in many parts of the developing world are already above 2 tonnes per capita.

Crucially, many developing countries are also growing quickly, and all have understandable aspirations for further rapid growth. It would be both morally wrong and politically impractical to propose that developing countries sacrifice their ambitions for growth and development.

In setting targets and policies, the world must recognise that poor countries will see emissions grow for some time, yet the richer among the developing nations will need to see emissions start to fall in around ten years if the 50% global cuts are to be achieved by 2050 (the majority of those nations will need to see emissions falling before 2030). However, it is not *slower* growth that will allow developing countries to achieve this fall in emissions – it is low-carbon growth, using technologies demonstrated and shared by the rich countries, as well as their own technological advances and drives for energy efficiency. We must find low-carbon growth for the world as a whole.

Developing countries should start planning credible action plans on this basis now. 'Credible' here means credible both to a country's own population and to its partners in a global deal. The targets should eventually include 2 tonnes per capita for approximate actual emissions by 2050, but paths between now and then could be specified in different ways – in terms of levels and rates of reduction of emissions per unit of output, for example.

Strategies have to place both mitigation of *and* adaptation to climate change at their core. In principle, one way of achieving this might be for an agreed global deal in Copenhagen to assign emissions targets to most countries, including major 'emerging' emitters such as Brazil, China, India and Indonesia, although the political feasibility of agreements over those targets, in terms of what is internally acceptable at this stage, is a major problem. The targets that these countries would see as commensurate with their needs and development goals would, in all likelihood and with existing technologies, be higher than 2 tonnes per capita. In China, for example, getting from the current emissions levels of 5 tonnes per capita to around 2 tonnes by 2050 would have to occur over a period when national income is expected to rise by a factor of ten to fifteen. To demand targets for actual emissions of around 2 tonnes per capita requires showing that it is consistent with development objectives. There must be substantial evidence both that low-carbon growth is feasible and that there will be substantial external technological and financial assistance along the way.

That evidence must be put in place quickly. Creating it through their actions would be in addition to rich countries taking on the 'at least' 80% cut by 2050 with credible interim targets. Beyond wanting to see that low-carbon economic growth is both achievable and affordable, it would be understandable if developing countries placed further conditions on rich countries before taking on caps themsleves. These could include a convincing demonstration by rich countries that financial flows to countries with cheap opportunities to abate greenhouse gases can be substantial, and that low-carbon technologies will be available and shared, allowing poorer countries to innovate, develop and ultimately export their own technologies.

Nevertheless, part of the credible action plans for middle-income countries, to be adopted within an agreement in 2009, should be conditional binding caps, which for many would take effect before or at 2020. These caps cannot be decided upon now, but would be subject to experience over the next decade, and would differ both according to local circumstances, and also reflect countries' 'common but differentiated responsibilities', including historical contributions to emissions. Once countries met certain 'graduation criteria' they would be expected to assume emission caps of the type adopted for industrialised countries, and would thus become involved in two-sided trading. The conditionality for these caps would be in relation to the performance of the rich countries in terms of their own reductions, financial flows and the demonstration and sharing of low-carbon technologies. Thus developing countries would 'commit to commit', subject to the performance criteria on rich countries and graduation criteria related to their own circumstances.

We are accustomed to rich countries and international financial institutions imposing conditions on the developing world in exchange for development finance. Expectations on the part of the developing world for actions by rich countries, in return for moving from one-sided trading to a cap for developing countries, is an example of this process in reverse.

There are a number of possible indicators that could come, in certain combinations, into the graduation criteria. An obvious one – which needs to be prominent, given the issue – should be emissions per capita. China, for example, is likely to reach the world average within ten years, possibly much sooner. It might take India twenty years, but again that could be sooner. Given that economic development is also a crucial objective, another set of criteria would be development indicators, including income and GDP per capita, and other metrics of growth. The classifications and indicators of the UN and the international financial institutions are valuable here. Adjustments for emissions from deforestation are also relevant. The issue is not simple, but definite criteria are necessary.

Targets should also take into account the difference between actual emissions in a country, and emissions associated with consumption and investment: the point here is that many countries 'import' a lot of emissions through trade. If a country sells financial services and buys cars, then the emissions involved in the construction of the cars which are imported can reasonably be seen as, in part, its responsibility. As the international division of labour shifts much manufacturing to the developing world, this has become an important issue in assessing the appropriate allocation of emissions. Allocation of rights to emit should take account of patterns of *both* production and consumption.

In approaching commitments to reduce emissions, the task should not be seen as a 'zero-sum game' of 'burden sharing' akin to carving up a cake. All too often, in my experience, international negotiations proceed in this way. Those who act early to find low-carbon growth will find many pay-offs for their own internal development as well as global markets for their innovations. There is so much that developing countries can do through energy efficiency and cleaner growth which will be in their own interests and of real value to their development programmes.

ONE-SIDED TRADING MECHANISMS

Developing countries will need to participate actively in the global carbon market. Prior to taking on binding caps this is likely to happen through an expansion of schemes such as the Clean Development Mechanism discussed in Chapter 6, in which we have a baseline-and-credit system where a firm can sell emissions reductions below a given level but is not penalised for going over it. Developing countries should come forward with their own design mechanisms for a reform of the CDM to make full use of trading opportunities. Interim schemes would be temporary, given that the eventual mechanism should be global cap-and-trade.

The CDM in its current form is not able to generate or absorb the financial and technological flows needed under a 'global deal'.[5] It has been estimated that climate stabilisation would imply annual global carbon flows through cap-and-trade of $20–75 billion by 2020, and up to $100 billion by 2030.[6] By comparison, the current CDM sees about four hundred project registrations per year, resulting in new financial flows of perhaps $6 billion at current carbon prices. The project-by-project nature of the CDM, and the measurement of emission reductions against an unobservable, project-specific, baseline representing business as usual impose substantial transaction costs in terms of validation, verification and independent scrutiny. From validation to registration, the CDM regulatory process currently takes about three hundred days, on average, and transaction costs can easily reach $500,000 per project.[7]

The parties to the Kyoto Protocol and the CDM executive board have recognised these problems, and various proposals to remedy the situation are under consideration. However, they may not be sufficient. Regulatory complexity and a high level of scrutiny are inherent in a project-based system such as the CDM. Since monitoring, verification and regulation are critical for their success, a different conceptual and institutional approach may have to be found to reduce transaction costs and achieve the necessary scaling up of financial flows.

Given that a target-based cap-and-trade is not appropriate for most developing countries in the immediate future, such an approach would still have to be based on the current baseline-and-credit system, but would have to move from a project-based system to a more wholesale approach. This could take the form of sector targets, or emission reductions associated with substantial programmes. These would probably be based on efficiency targets, rather than sector caps, although sector caps may be possible for globalised industries such as steel. For example, for each tonne of cement produced using less than an agreed amount of carbon, producers in a developing country could sell the difference as credits. Similarly, a large power utility might sell the sector-wide emission reductions from a reform programme that includes energy efficiency and the roll-out of technology for renewables like wind or solar, instead of coal-fired power stations.

Consistent with the sustainable development objective of the CDM, the benchmarks could include social and environmental sustainability criteria, for example in terms of local environmental protection or environmental, health and safety standards. Care should be taken, however, that sensible safeguards do not involve vexatious bureaucracy and arbitrary hurdles. To maintain the incentive to innovate and avoid excessive rents, benchmarks would probably need to be strengthened over time as new abatement opportunities become available.

Defining benchmarks in practice would not be without difficulties. There is a need for standardised efficiency factors to allow a move away from the administratively costly case-by-case argumentation of the current CDM. But there also has to be recognition of local circumstances. Low-wage countries, for example, are likely to adopt more labour-intensive production processes, while resource endowments like wind regimes and water reserves will determine the local fuel mix. Similarly, different benchmarks may have to be adopted for different products or production processes – integrated steelworks and 'mini mills', for example. If carbon-intensive activities can be outsourced (as with cement), the entire supply chain may have to be considered. Other problems might arise from data limitations, and the potential unwillingness, for competitiveness reasons, of companies and countries to share data. It will be important to avoid overcomplicating the criteria.

Standardised benchmarks may therefore not be possible everywhere, and in some sectors the current project-by-project approach might best continue. However, sectors where a wholesale approach might work include most emissions and energy-intensive industries, including electric power, refining, pulp and paper, metals and cement. Sector benchmarks may also be a good way of incorporating international transport emissions (airlines, shipping) into the global deal. More work is needed, though, to find a solution for sectors with emissions from a wide range of sources, such as agriculture. Another sector that may deserve special attention is land use, land-use change and forestry. In particular, the balance between afforestation or reforestation (included in the CDM), and the rewards for avoiding deforestation (not included in the CDM) needs further thought. At present there are anomalous incentives in current trading arrangements – you can, in principle, cut or burn down trees without penalty or forgone reward, and then get CDM assistance for reforestation.

In sectors that are particularly subject to international competition, such as aluminium and steel, the benchmarks would probably mirror the efficiency levels expected from firms in industrialised countries (for example, those used in the allocation of allowances in an emissions trading scheme such as in the EU). This might take the form of global-sector agreements. Standardised benchmarks would help to reduce the risk of carbon leakage, alleviate competitiveness concerns, and thereby help to preserve free trade.

The one-sided nature of the proposed mechanism makes it possible to set fairly stringent benchmarks. Similarly, the mechanism would have to cover a broad set of acceptable technologies, including, for example, CCS. Both sector and technological benchmarks which cover the full range of options are important to facilitate developing nations, particularly middle-income countries, participating actively in the global abatement effort.

It is possible that all this detail on amending the CDM has bored some readers and left others hungry for more so that they could see how it might really work. Our purpose here is not to provide a full programme of reform; it is to provide a clear sense of direction and indicate that the necessary changes are both feasible and valuable. For those for whom the detail is excessive, there is a message: detail matters, and there will be a lot of hard work to get it right. This in turn requires both an understanding of the practical realities of market places and technologies, and a clear-headed application of basic principles.

3. International emissions trading

The essential reason why international emissions trading should be a key element of a global deal is that it embodies the three principles on which the deal as a whole must itself be founded. First, a cap imposes an absolute limit on emissions, and therefore clarity on reductions; thus it gives effectiveness. Second, competition and the market will seek out the cheapest ways of reducing emissions; thus it gives efficiency. Third, the structure of quotas, together with the exploitation of some low-cost emission reductions options in developing countries, can generate private-sector finance to developing countries to support low-carbon growth; thus contributing to equity. These financial flows can provide part of the 'glue' for a global deal. For the reasons described in Chapter

6, it would be very hard to meet these criteria with a system that relied only on national taxation and regulation, important though these will be.

The building of international markets should – and this is a crucial task – cover the *linking* of rich countries' trading schemes (such as the EU ETS), since they will initially develop most strongly, because the rich countries have already taken on explicit capping obligations under the Kyoto Protocol and should be leading the way on caps and reductions in the post-Kyoto arrangements.

It is probable that in 2009 or 2010 the USA will enact a cap-and-trade bill. Cap-and-trade schemes in Australia and New Zealand are likely soon. One feature which could cause problems for linking, and on which there has been much discussion, concerns price ceilings or 'safety valves'. The latter term is somewhat misleading, however, because they are not safety valves from the perspective of the dangers of climate change. Indeed, they are the opposite: if prices from a given cap turn out to be high, they allow the cap to be broken by, essentially, the government selling as many permits as necessary to keep the price down. Further, these types of caps limit the potential for trade and thus promote international inefficiency: if the price is kept artificially low on one market, then everyone will attempt to buy there, and at a price which does not reflect marginal costs at the desired overall cap; hence there would have to be restrictions on trade.

Sometimes, on the other hand, the worry is that the price will be too low if international trading is opened up and thus domestic incentives to cut back will be blunted. The answer to this question, surely, is not to curtail the opportunity to lower the costs of reductions via trade, but to raise the ambitions and get more reductions for the money. Precisely because of this fear, the EU has been discussing curtailing the amount of CDMs that can be bought from developing countries in order to meet reduction targets of 20% by 2020.[8] The answer is to commit to 30% reductions and to more trade. It would not only achieve more, it makes it more likely that other countries will set their sights higher as part of a global deal. At present the EU has indeed indicated, as part of a global deal, a move to 30% cuts by 2030, and greater openness to trade.

There are many definitions and rules in carbon markets which could complicate the development of trade. The importance in keeping costs of action down, and the role of trade in doing so, are such that early work on the design of schemes should give a high priority to the avoidance of apparently minor definitional issues, which could lead to major practical barriers if they are not spotted in advance.

Early work from the 'GLOCAF' model of the potential for world trading (being developed in the Office of Climate Change in the UK) suggests that globally, and relative to a world trying to bring about reductions without trading, costs of achieving targets could be halved by 2020 (and possibly up to 80% lower) with full cap-and-trade. Gains for 2050 would be smaller because by then most countries will, we hope, have taken significant action and there will be fewer cheap abatement opportunities to sell. In this sense we would expect the volume of trade to rise over the next twenty years or so, and then start to fall. That would be a feature of success.

One feature of emissions trading which has attracted considerable attention is how to treat imports from countries which are not making sufficiently serious (or perceived as such) efforts to cut emissions. Here such countries are seen as acquiring an unfair cost advantage by failing to take on the extra costs that more 'responsible' countries and firms are taking. This is one of those cases where a countervailing tariff might in logic be justified. The magnitude of these cost advantages is, however, likely to be small – in most cases of the order of 1–2% of the import price, or often considerably less. Also, the evidence from studies of the mobility of firms in response to environmental policies is that it is negligible.[9]

All too often the plea for the countervailing tariff or measure is a slogan without numbers or hard evidence (actually covert protectionism). Four decades of working on international development and examining the real gains from the openness of economies has made me deeply sceptical about such attempts to distort trade. While at a theoretical level the argument for a countervailing tariff is not wrong, in the case of cap-and-trade the competitiveness impacts of such a scheme are of small importance and affect just a few industries. However, the impacts are usually, implicitly or explicitly, so exaggerated as to suggest that the justifications used are protectionism in disguise.

For the few sectors which use energy very intensively, such as aluminium, steel, cement, paper and refining, there may be a case for some transitionary arrangements. The industries themselves should play a prominent role in working these out. It is very important, however, not to let the 'tail' of a few energy-intensive industries, wag the 'dog', which is the quick development of a large and well-functioning international carbon market. Nevertheless, there is no doubt that the competitiveness issue concerning measures like cap-and-trade, or mitigation more generally, looms large politically – as we see in the next chapter.

All of this discussion of the detail of cap-and-trade schemes (and much important detail has been omitted) may lead the reader to ask again whether there is a simpler way, in particular an international carbon tax. I do not want to repeat the arguments of Chapter 6 on taxation and cap-and-trade. Within many countries taxation can and should play a major role. But it cannot give the same confidence over long periods of stability of policy, of actual reductions, and of financial flows to developing countries. And we do not know in a world of oligopoly in oil and gas markets what the effects of tax on prices will be. Trading must play a central role.

4. Deforestation

The whole world gains if a country prevents deforestation, stops forest degradation or undertakes forestation. Changing land use – largely deforestation, peat burning and the like – is responsible for close to 20% of current emissions, similar to the total from the USA. The World Bank has recently estimated that more than 1.6 billion people, a quarter of the world's population, 'depend to varying degrees on forests for their livelihood'.[10] It is they, and the countries in which they live, who understand their goals and the structure of their economies, societies and natural endowments. It is for countries themselves to design their own development strategies, including the ways in which the relationship between forest and development is managed, but there is much that others can do to help in terms of sharing experience and ideas so that any particular plan is as well informed as possible about opportunities and risks. The financing of pilot projects around the world will be crucial in finding ways forward which are effective in reducing deforestation and enhancing forests while at the same time being harmonious for development. In this process, international

institutions such as the UN and the World Bank have an important role
to play.

A difficult, yet vital, task, however, is to combine these learning
processes with speedy international action. We are losing 13 million
hectares of tropical forest a year, an area half the size of the UK. And
unless we act globally, a particular action in one place, effective though
it may appear locally, could simply shift deforestation somewhere else
(this is sometimes called 'leakage'). Estimates of the costs involved in
halving deforestation range from $3 billion to $33 billion per annum.[11]
The more action is taken on a global scale, the less leakage: some of the
higher estimates come from pessimistic assumptions on leakage (if it is
assumed to be 50%, for example, then the cost of a given amount of
reduced deforestation is doubled).

This is an area where the world can move ahead quickly and get
results soon, at reasonable cost. We should commit – and this could be
part of the Copenhagen deal in December 2009 – to at least $15 billion
of public funds per annum, with the aim of halving deforestation
within a few years. The important Eliasch Review[12] has provided the
most recent careful study of such a programme and gives estimates of
costs of halving tropical deforestation in the range of $15 billion per
year. Initially, it is likely that these funds will have to be public because
of the centrality of basic development issues to much of the action. It
may cost a bit more or a bit less (this figure assumes that action is
globally coordinated so that leakage is under some control); we can
find out only by taking on the problem at scale. To halve deforestation
would save around 3 Gt CO_2e per annum, and thus the world would be
'buying CO_2' at around $5 per tonne. That is a very good deal
compared, say, with 2008 prices in the European Emissions Trading
Scheme of around $30 per tonne. It should be a priority from the point
of view of climate change, and has a whole host of associated benefits
in terms of biodiversity, water control and development.[13]

The policy response depends largely on which of the several reasons
for deforestation prevails in a particular region. Where some of it arises
from very land-intensive methods of agriculture, such as slash-and-
burn, it can be reduced by encouraging techniques with much higher
output and labour per hectare; where it arises from the need for wood
or charcoal, alternative fuels need to be made available. In some cases
the primary cause will be international demand for commodities such

as palm oil or soya beans. We should ask whether alternatives to these commodities could be developed which have less need of good land and water. Moving quickly to second-generation biofuels will have an important contribution here. The knock-on effects of the current generation of biofuels, corn and sugar-based ethanol and palm oil, are very worrying in terms of the effects on food prices arising from displacement, directly or indirectly, of land which could be used for food, or particularly for palm oil, in terms of deforestation. Some of these examples, such as corn-based ethanol, may even be making a negative contribution to reducing emissions.[14]

How can we ensure that the full value of the forests comes into both public and private decision-making – in other words, how do we foster a demand for forests and for land for other uses, particularly for food and biofuels, and coordinate action?

The forest and the land on which it stands should be put to the use with the highest social value, according to valuations which take proper account of all the impacts. Often the alternative use of the land has direct 'use value' of just a few dollars per hectare per annum, whereas the value as forest is far higher. Looking carefully at these values usually suggests that preserving forests has a much greater social productivity than cutting and burning (which is the cheapest way, without a carbon price, to clear) in order to grow alternative products. Such calculations can be done in terms of the social value of forest as forest, versus the social value of the forest in alternative use, taking into account costs of the transformation and the value of wood obtained. It is estimated that the Amazon rainforest derives more than 50% of its social value from non-timber usage, such as erosion control, absorption and storage of CO_2, or simple existence value (the satisfaction people have from knowing it has not been destroyed).[15] The value of leaving the trees standing should be compared to that of cutting them down. If the analysis is done properly the former will, for tropical forests, generally be much higher. Put crudely, the forest is worth much more standing than cut.

In asking about policy, however, there are of course complexities to be addressed, such as the enforcement costs of preservation and the incentive structures necessary to lead private individuals or enterprises, local authorities and communities to choose preservation in the first place. Incentive structures require a mechanism for rewards for keeping the forest which corresponds to the excess of the social value of the forest above private returns from its alternative use; in other words, policy creates a monetary demand related to the social value which is created by protecting the forests. This involves a potential reward – usually a direct payment – related to the value of the carbon, the biodiversity, the water impacts and so on, as well as penalties for those who undermine preservation efforts. To create and maintain the necessary incentive and enforcement structures involves real costs and the building of administrative and institutional capacity and outside support is required. Much of the organisation has to take place at the national and regional level since it has to do with overall governance, legal and enforcement issues, and the specifics of policy depend on the way in which forest and near-forest communities function, as well as on the overall development objectives and challenges faced by the country. If it turns into a contest between the forests on the one hand, and economic growth on the other, or between the country where the trees stand and the outside world, then the forests will lose. For the planet and our future, we cannot afford to let that happen.

These issues make it clear that the necessary measures are closely intertwined with the promotion and funding of governance and 'soft infrastructure' aspects of development. That is why, in the first stages, establishing the flow of resources to national and local governments is the biggest priority. This also has the advantage of taking account of leakage within the country, in that financial rewards would be linked to overall country performance. International leakage would be reduced as far as possible by making action global.

Over time it should be possible to blend these with private flows – indeed, such possibilities should be explored from the beginning. Whether the flows are public or private, they should be linked to results. In other words, a government in a country where trees stand, be it central, regional or local, would agree a contract to achieve certain results and the flows would be linked to those results. Some international trading schemes, if suitably structured, could be developed to allow private flows to provide incentives for emissions reduction in this way.

Looking further ahead, these schemes should plan for more direct integration into emissions trading. As these systems are built and the governance and 'soft infrastructure' strengthen, then it should be possible to integrate trading between local forest-related entities and firms inside or outside the country seeking to buy emissions reductions. At the early stages of global trading schemes it is important that prices are not depressed by a very big cheap source, while at the same time it is vital for keeping down the cost and sustaining action to exploit the cheapest sources first. In this case bringing deforestation into trading on a major scale is several years away, as much of the initial cost of action involves investment in governance, and public soft and hard infrastructure for the economy as a whole as well as for the specifics of deforestation.

Although developing an effective, efficient and equitable policy on combating deforestation and enhancing forests poses difficult problems, both of strategy design and implementation, the outlines of where we need to go are fairly clear and the task is urgent. There should be an early commitment of resources for the task, of the order of $15 billion, to promote immediate investment from public authorities and private investors in the mechanisms and policies required to secure the financial viability of the tropical forests in the long term. This should be backed by a credible, long-term funding commitment from the industrialised countries as part of a new global deal, which could combine commitments to gradually liberalise access to carbon markets in the medium term, with direct funding to build capacity in the short term. If that commitment is made and the outlines are agreed, then effective action is likely to develop very quickly. The countries where the trees stand are very keen to take on the challenge of combining development and preserving forests – witness the offer of the president of Guyana to have international management of Guyana's forests, as part of an international scheme and of Guyana's development.[16] The Coalition of Rainforest Nations has many promising ideas, among them how to reduce deforestation through tapping into the emerging global carbon market in such a way that would not destabilise the overall market. The World Bank is making it a priority. International resources and commitment on this scale would produce a strong response. We must be clear, however, that halting deforestation and promoting economic development can and must go hand-in-hand and the design of action on both is for the countries and communities where the trees stand.

5. Technology

Cutting emissions to the required level will need rapid and widespread advance in the development and diffusion of technologies. Well-developed carbon markets, taxation, regulatory and other policies will create a powerful stimulus, as discussed in Chapter 6, along with specific policies to promote technologies. The challenge in this chapter is to indicate how the necessary diffusion and advances in technology can work on an international scale as part of a global deal. The crucial importance of a price for greenhouse gases is taken for granted in what follows, as it is absolutely fundamental for the right incentives, although many options will have great attraction simply from the perspective of energy efficiency.

In Chapter 6 three different (though related) tasks, with varying time horizons, were discussed. The first was diffusing existing low-carbon technologies, the second was scaling up new commercial technologies and the third was creating breakthrough technologies. The global aspects of these tasks or associated horizons are different, with three broad international instruments for action.

The first instrument is globally coordinated standards, and openness of markets to trade to encourage fast deployment of existing technologies. In particular, this would allow us to make the most of existing technologies for improved energy efficiency. Important examples include electrical goods, buildings and transport. In part this is to allow a level playing field for competition, and also to allow for economies of scale.

Unleaded petrol and catalytic converters have become almost universal as a result of standards. In the early years of this decade, auto-rickshaws and diesel-engine buses in Delhi were required to switch to cleaner fuels, particularly compressed natural gas, which has resulted in a dramatic effect on pollution. The world is now learning from Sweden the value of ground-source heat pumps in heating buildings, where they have been used successfully for thirty years. Electrical appliances which turn themselves off after a period on stand-by can save a great deal of electricity. These are examples where appropriate regulation and an international perspective can separately, or, even better, together, yield strong results. Procurement of equipment and infrastructure for cities around the world could follow common climate change standards, allowing for the development of demand for hardware on a major scale.

The second instrument is coordinated public funding for demonstrating and deploying critical technologies in both developed and developing countries. Important examples here include CCS for coal and second-generation biofuels. The challenge of reducing emissions by 2050 will be greatly compounded if CCS for coal proves very problematic or costly. The only convincing way to discover its potential is to build around thirty commercial (above, say, 300 MW) plants in the next ten years.[17] The discovery process will need many plants because local coals and geologies vary, as do legal structures. Current calculations suggest that for a total price support of $5 billion per annum,[18] such plants would be forthcoming from the private sector. The support would be temporary if strong carbon trading replaces it. However, the risks from the experimentation required for initial plants, and the gains to others from learning-by-doing and learning-by-watching, justify public support, which should be coordinated internationally because we all learn from the experience in one place or country. One way of providing support in the case of CCS would be for a group of countries to take on the task of building a given number. For the UK, as a rich country with around 2% of world emissions, to fund two or three plants would seem a reasonable contribution; the current plan for the UK of just one looks too few in relation to the thirty needed soon, and the international responsibility of a wealthy country that has been emitting for a long time. If Europe took on a dozen, the USA something similar, Australia two or three, and so on, we would soon get there. The supportive instruments could vary in different countries, and that too would be a valuable piece of learning.

The problems of CCS are not just financial and technological, they are also legal and regulatory. Who owns the underground chambers where CO_2 might be stored? Who is responsible for leakage in storage and transport? Responding to these questions is part of the learning process which we have to embark on soon.

I have emphasised CCS for coal since it is quantitatively such an important challenge, given that China and India are likely to use 70–80% coal as they expand their electricity generation, and coal will be widely used in many other countries. We must learn quickly whether it will succeed or fail. Even though it may be only a transitory technology between 2020 and 2050 as other technologies are developed, that is a crucial period.

Biofuels constitute a second very important case. Although it is likely that electric or hydrogen road transport will play prominent roles in a few decades, biofuels may well be a significant part of supply between now and then. And if serious advances are made in second-generation biofuels (which have much lower requirements for water and good land), then they may compete favourably with electricity or hydrogen over the longer term. That will be determined in large measure through the markets, although possible effects on international food supply and prices would be a matter for international concern and action.

One can never be sure how options and preferred technologies will develop. That is the nature of discovery and markets. Nevertheless, given likely demands for fuel, for a few decades, in road transport (and probably longer in shipping and aircraft), it makes sense to research the future of biofuels very intensively, especially because of the problems associated with them in their current form. Part of the research could be in internationally funded research institutions like the Consultative Group for International Agricultural Research, which has long experience from the Green Revolution of the 1960s onwards in developing agricultural crops and techniques suitable for environmental, social and economic conditions in developing countries. Over more than thirty years of research in the rural Indian village of Palanpur in Uttar Pradesh, I have seen directly not only the impact on output per hectare and per worker that new varieties and techniques can have, but also the importance of examining very carefully the potential social and environmental impacts. While output per hectare has risen strongly in Palanpur the water table has dropped from three or four metres when we first visited in 1974 to over ten now. Other parts of the research would have to be in close coordination with technologists from the car, shipping and aeroplane industries.

Design of the infrastructure for the delivery of alternative fuels raises a further class of issues. But this can happen quickly: visiting a Brazilian petrol station is like going to a bar – there is a whole range of taps available for different types of fuel. Public-private partnerships will be crucial in this whole area, both in the technologies themselves and in the supporting framework using them. Part of the quid pro quo for public support would be programmes for rapid sharing of results.

A further aspect of research should be the effects of biofuels on global land, food and water markets (not the type of issues normally examined by the R&D departments of major firms). Once we begin to look at the work in detail, it is clear that for key aspects a structured world programme is necessary. Part of the research will be driven by uncoordinated R&D departments of firms working in competition and with each other, and the competitive market place is a crucial stimulant to creativity; other parts will be by international institutions looking at global effects; and others by public policy analysts working for national governments, international organisations and so on. We must take a very broad view of appropriate structures of R&D – some should be managed and coordinated, others not.

Agriculture poses its own special sets of problems, since issues of adaptation, mitigation and development are closely intertwined. Research should focus on techniques which link and embrace all these areas, such as those which foster a greater resilience to adverse conditions and climate change (and thus make sense for adaptation), on techniques that release less CO_2 and other greenhouse gases, and those that call for less tilling and less water, raising land, labour and water productivity and thus boosting incomes and development.

Nuclear fusion may require international collaboration on a major scale because of the size of the necessary investments in experimentation, the very specialist skills (which take a long time to develop) and the global risks that may be involved. Indeed, R&D in this area may be dominated by a few big international centres for research.

The organisational mechanisms would have to be very varied. Speed, incentives and the absence of bureaucracy would be of the essence, features not always associated with international structures. Private-sector entrepreneurship, based on competition, would be a key driver. International intellectual property rights would have to be managed in a way that combines incentives and openness.

All this makes it clear that the R&D challenge requires the funding and the powerful commitment of a Manhattan Project, or the moon landings in the 1960s, both of which showed that intense focus, high levels of skill and strong resources can produce new and major results very quickly.

However, while in terms of priority, speed, resources and skills the Manhattan analogy makes sense, in other ways it is misleading. The structures, funding sources and mechanisms we need now are very different from those associated with a single-minded group of scientists and engineers, in just one country, with a limited range of skills and focused on a very narrowly defined objective. This challenge is far more international, much more subtle and diverse in its definitions and subjects, more dependent on the behaviour of people and firms, and more intertwined with the many problems of development, water, biodiversity, security and so on that the world faces.

6. Adaptation

The next forty or fifty years of climate change are shaped by what we have already done and what we will do over the next decade or two. The developing world will be hit earliest and hardest, while the rich world has more resources and technologies, and the majority of the responsibility for past emissions that have taken us to the very difficult starting point for action. The sense of injustice felt intensely and very understandably by the developing world is a crucial element in the perception of the equity and thus the feasibility of any global deal. A major effort by rich countries to support the adaptation of developing nations is a key part of an equitable deal.

In most societies, wealth brings responsibilities. Notions of the rights of others and liberty from interference ('negative liberty', in the language of philosophers such as Isaiah Berlin, John Locke and Thomas Hobbes) are crucial aspects of responsibility for the consequences of our actions: emissions interfere with the rights and liberties of other people. Long-term incentive structures to promote due care and attention to the effects of our actions is an area which economists examine. More detail on the challenges of adaptation was given in Chapter 4; the focus here is on responsible international action as part of a global deal.

As with other issues such as deforestation, policies for adaptation should be set by developing countries themselves as part of their development programmes. It makes no sense to formulate these programmes as if climate in the future will be like climate in the past. At the same time it makes little sense to try to define adaptation policies and necessary resources separately for each project and programme. Such calculations would be diversionary and a waste of scarce time and resources. The challenge is to get the best possible information on how the climate is changing or might change, and plan activities and investments accordingly.

Just as most development funding will have to be internally generated, this will also be the case for the extra costs resulting from a more hostile climate. It will be the farmer, the firm and the local community that will have to adjust techniques, crops, irrigation, buildings, transport, flood control and other forms of infrastructure. The extra cost is, however, relevant to a world that should recognise some collective responsibility, both for international development goals and for the difficulties caused by climate change.

Estimates of overall costs of adapting to climate change are in their early stages. While the arguments set out in Chapter 4 suggest that a search for great detail in calculating additional costs from climate change for all sectors, techniques and so on, is unlikely to be very useful, broad orders of magnitude are relevant. One such overall view was taken recently by the UNDP in their Human Development Report 2007–8. Looking at the challenges of investment, of human development, and of social support associated with development objectives, they suggested, as we saw in Chapter 4, a figure of around $85 billion per annum by 2015 as representing the extra costs arising from a more hostile climate. On the basis of a rather narrower definition, the UNFCCC indicated $28–68 billion per annum by 2030. I find the broad approach of the UNDP in relation to impacts and adaptation persuasive. But what is very clear is that these costs are of the order of scores of billions of dollars per annum in the next decade or two. They would rise very rapidly into the many trillions if climate change is not managed sensibly. As a working hypothesis for the purpose of illustrating the scale of requirements, I would suggest a range for the scale of funding of $50–100 billion per annum for 2015, rising substantially beyond that as we move into the following decade.

How should international support and funding for adaptation be organised in a way which is integrated with, and not disruptive to, programmes for development, and how could external funding on the necessary scale be generated? Two examples of potential ways forward for organising support from multilateral institutions are, first, for a new 'adaptation funding window' at the World Bank which works alongside its International Development Association (IDA) 'soft window'.

The IDA window provides funding on a basis which is effectively 70–80% grant. Generally, IDA has a ten-year grace period followed by very low interest rates and long periods before principal repayments begin. It is usually allocated across countries with a strong focus on two criteria: first low per capita income, or a high incidence of poverty and second, a productive environment for the use of resources assessed in terms of policies and government capacity. It has its own governance structure within the World Bank Group. A new funding window could be organised in a similar way. It could, and arguably should, be grant rather than soft loan, and of a similar magnitude. I would suggest $10 billion per annum initially. Clearly that would be small in relation to requirements, but there should be other international windows too, through the UN and its associated Environment Programme and Population Fund, for example. Different institutions and funds would have different priorities and call on different skills. If the two windows are side by side within an institution, there should be few real problems of integration.

A second suggestion is to scale up IDA allocations in a way that takes into account the adaptation challenges and risks country by country. However, there are problems with this approach: it might politically be treated by some countries as giving scope to future attempts to avoid increasing resources on the scale required for development in a more hostile climate, because the discussions might 'look like' those for standard 'IDA' replenishments (which take place every three years or so and provide funding for soft loans by the World Bank to the poorest countries); and it would be likely to miss an opportunity to generate further governance reform of the World Bank.

While, of course, these are not the only two possibilities, they do illustrate ways of providing external support which try organisationally to take account of the importance of country leadership in policy, of integration with development strategies, and of potential scale. Other routes could include the regional development banks, such as the African Development Bank, the Asian Development Bank and the European Bank for Reconstruction and Development. Each has its own strengths. And an element of competition between institutions as well as collaboration does no harm. There should not be a single window or source in just one institution since all development institutions must take climate change seriously. Coordination is a very important issue, as was recognised by the Paris Declaration on Aid Effectiveness,[19] endorsed in 2005, and which provides very useful guidelines for ensuring assistance from multiple sources is delivered both effectively and efficiently.

A new IDA-like window has potential leverage for further governance reform of the international financial institutions. The Bretton Woods structures were designed in 1946, in a world recovering from the Second World War and the Great Depression and before decolonisation, and were dominated by a few rich countries. The governance structure has not altered in a way that it is at all appropriate to the changes in the world since then. This new window, conceived when we have now realised the significance of the threat to humankind, would be a clear example where the international community could choose a governance structure reflecting the world in the twenty-first century. The Global Fund for HIV/Aids, Malaria and Tuberculosis is a recent example of innovative organisation which could be examined for lessons. Founded in 2002, it instituted a system based on competition and payment by results – an incentive structure for delivery – that appears to have worked well.

Given the necessary integration into development funding and historical responsibilities, we have advanced powerful arguments for including adaptation as part of development funding. Some might understandably ask whether funding for adaptation would in those circumstances be additional to existing commitments. From one perspective the question is misguided – adaptation is not a task separate from development, that may or may not be tackled depending on whether additional funding is available. The question should be about the overall level of development funding in the context of a climate which is more hostile than was fully understood when the world set the Millennium Development Goals (MDGs) at the turn of the century. The funding necessary to achieve the MDGs is higher than was understood in 2002, when the UN Monterrey Conference discussed requirements. And it is higher than was anticipated when the report of the Commission for Africa was presented to the G8 Gleneagles Summit in 2005. Since I was at Monterrey as the chief economist of the World Bank, and I led the writing of the report of the Commission for Africa, I am all too clearly aware that climate change had a relatively low profile on both occasions.

The implication of giving climate change the prominence we now understand it to require is that we should go back to our estimates of the support that is necessary for the MDGs. From this perspective, by 2015, it is likely that new calculations would be of the order of $50–100 billion per annum higher than was originally thought. If this is roughly right, then where should we go from here? My response is in two parts: first, the rich countries should deliver on the Monterrey and Gleneagles promises; they are currently way behind. If the world achieved 0.7% of rich-country GDP in aid by 2015 (this is the EU promise), aid would be around $300 billion per annum compared with around $100 billion now (and we are at $100 billion now only with fairly 'flexible' accounting methods that give strong weight to debt relief). The extra resources of up to $200 billion, or more, from such delivery would make a major contribution to the extra challenges posed by adaptation over the next five to ten years. So let the rich countries start with delivery on their promises.

Second, looking forward, we will soon have to ask questions about development goals and support beyond 2015. As we do this, we must not make the mistake of placing climate change anywhere other than centre stage. There is no doubt whatsoever that the necessary funding as a result of the extra challenges posed for development by climate change will be high. As has been a constant theme in this book, if we do not tackle climate change and development together, we succeed on neither.

Funding sources

The first three elements of our suggested global deal concern targets and trading. The second three – deforestation, technology and adaptation – all involve funding. For international public funding by rich countries I have suggested, as a start, the following during the next decade:

- $15 billion per annum to cut deforestation in half.
- Public energy R&D research to increase from $10 billion to around $50 billion per annum (i.e. an extra $40 billion). This is part national and part international, but for completeness we include it in the global deal.

- Around $75 billion per annum for adaptation in developing countries (taking the centre point of the range $50–100 billion).

Taking these three elements together and counting fully the R&D expansion, the total international aspect of public funding is around $130 billion per annum, or around 0.3% of rich-country GDP. That is a fairly modest sum in relation to the enormous gains to all countries. The private flows in the carbon markets from rich to poor countries could be of the order of $100 billion during the 2020s. As in all serious policy discussion, we must have a clear understanding of the 'bill': this is the rich-country bill for the international part of the story.

The national costs of meeting their own targets might be of the order of 2% of GDP most of which would come directly from private consumption and investment rather than passing through public budgets. The international element might represent around a quarter of the total cost (taking public flows and carbon flows together) and much of it forms part of the 2%.

What are the potential sources of the international public funding for a global deal? The broad answer must be the overall public revenues, principally taxation. That may not be great news for the finance ministers of the rich world, but these are expenditures from the public purse that make eminent sense. To say we cannot afford it is nonsense. Public revenues in rich countries are usually 30–40% of national income: 0.3% from the 30–40% (representing around 1% of government expenditure) for a global deal is money well spent. The returns in terms of climate security compare very favourably, in my view, with security benefits provided by defence budgets, which typically run at ten times this figure, around 10% of government expenditure in rich countries.[20] An expenditure on climate security as a global threat, of one-tenth of that on security from external threats from other nations does not look excessive given the dangers. The claim 'we cannot afford it' is not very different from 'we are not sufficiently bothered to deal seriously with climate change'. That is simply reckless.

On climate-related new sources of revenue, such as carbon taxation or revenues from auctioning of permits, we must be careful since they are part of overall public resources and, as with most revenues, can go into increased expenditure (on education or health, for example), lower borrowing or less taxation. But political earmarking of some of the revenue for international action on climate change might make both the global deal and the auctioning more politically acceptable. If, for example, permitted global emissions were around 30 Gt CO_2e in 2030, and half of this were auctioned, then, with a price of CO_2 around $50 per tonne, the total revenue would be $750 billion per annum. Perhaps less than half of the auctioned permits might be in rich countries, so revenue in wealthy nations might be in the region of $300 billion. That would be at least enough to fund the international expenditures indicated, although we must remember that substantial national public expenditures would be necessary too. Carbon tax or auction revenues do not provide an unlimited pot, but they are likely to be of sufficient order to more than cover international funding requirements for a global deal. In the case of adaptation funding, they could provide one element of a predictable flow of resources.

What I hope to have shown in this chapter is that the outline of an effective, efficient and equitable global deal can be described clearly. It is feasible and it can be financed. While there is much work to do on detail, and all sorts of variants can and should be constructed and examined, we know enough to get on with the job. We will find out more along the way.

CHAPTER 9

Building and sustaining action

Any global deal must be designed with a keen eye not only on how a very broad-based agreement can be built, but also on how it can be sustained. This is one of the reasons why the three principles of effectiveness, efficiency and equity are so important as the foundation and guiding ideas for a deal – if any one of these principles is badly and consistently violated, consensus will be very difficult to build and sustain. But there is much more to building and sustaining a deal than good design structure.

Fostering consensus across countries is an intensely political process in which self-interest, misperceptions and mischief can play powerful roles. Misleading lines of argument will have to be countered with analytical skill, and policies involving dislocation will have to be sensitive to genuine and understandable concerns. Similar talents will be required to sustain an agreement through the ups and downs of the international economy and changing political leaderships, electoral cycles and sentiments around the world. Both within countries and across countries, the international community must work to create a spirit of collaboration which is deeper and broader than we have ever seen.

The practical politics of different countries vary enormously, as do the challenges of persuading and building internal coalitions. Being able to show clearly what might be involved will be crucial – analysis matters. As part of these challenges there will be particular arguments that have to be dealt with specifically and directly, such as impacts on competitiveness in the rich world, or how low-carbon development can actually be transformed into a reality in the developing world.

Sustaining world action will require much of the leadership and many of the skills that are required to build it. Of special and central importance is demonstrating the principle of international collaboration in action. Cooperation, accommodation and compromise have been built with some success over the years in the EU, a tremendous achievement of integration and peace-building in post-war Europe. It may often be complex and messy, but with experience the collaborative spirit can be created, decisions made and results achieved. The leading role of the EU in making commitments on climate change is itself an example of European collaboration. So too is the EU further commitment to the building of a global deal.

Collaboration can also be helped by strong movements and organisations from the bottom up which sustain longer-term views and values. Institutions like the UK's Climate Change Committee can be constructed which act and speak up for longer-term goals. Such institutions will be necessary fairly quickly at the national level, and will be required internationally over the medium term. Keeping a long-term perspective through short-term disruptions will be a challenge for leadership.

Raising sights, understanding collaboration

The best of all foundations for a common vision for action is a unity in understanding that, first, climate change and global poverty are the greatest challenges of the twenty-first century, second, that we can, if we collaborate, respond strongly and effectively to both, and third, that we shall succeed or fail on them together; to tackle only one is to undermine the other. The arguments supporting that shared understanding are, in my view, very powerful and this book is one attempt to assemble and present them in a systematic way. But it will take commitment and communication from this generation of world leaders to translate argument and analysis into effective action.

Mahatma Gandhi shaped a non-violent movement across a diverse subcontinent to bring about India's independence with a clarity of vision and extraordinary powers of communication. There was no historical necessity why that movement became non-violent, and he set an outstanding example with profound implications, not just for post-independence India, but also for the understanding of decolonisation generally. While not necessarily a model for other countries, the process gave a perspective of what could be achieved, which was of world importance. Nelson Mandela's vision of post-apartheid South

Africa and his understanding of the vital roles played by collaboration and reconciliation brought an extraordinary spirit of cooperation to a nation that had seemed hopelessly divided. His powers of communication gave the first years of the new South Africa a vigour and stability in circumstances that might have led to conflagration. There was no inevitability about this process, either. Again, it was the vision of a remarkable man.

Of course, Gandhi and Mandela were not alone – there were remarkable people working with them, and key organisational structures were created: the India National Congress and the African National Congress. The two movements and countries learned something from each other. Extraordinary collaboration can be shaped and built by great ideas, powerful communication, visionary leaders and the right kind of organisation.

The challenge of climate change requires the same spirit of internationalism that founded the UN, and it requires the vision, communication and organisation of Gandhi and Mandela. Many of the details of the global collaboration will be prosaic, but the finer points of detail matter, and low-profile hard work will be of the essence. If sights are raised and problems understood, ways can be changed and new alliances built. All kinds of things can and will go wrong: the subcontinent suffered the tragedies of partition, South Africa is still a deeply unequal society and the UN is hardly a perfect institution. But they show us what leadership, vision and communication can achieve in responding to threats (and the opportunities they present).

One of the key elements in any collaboration must be an understanding of what others are doing. There is often a real suspicion among people of a particular country that while they are being called to action, others will do nothing. Information is crucial here. In the US, I have often spent time describing the actions China is taking to halt deforestation and to reforest, to promote energy efficiency, to tax energy-intensive exports and to impose strong emissions standards on cars, while also recognising its construction of one or two substantial coal-fired power stations a week and a very rapid growth of emissions. In China, I have constantly emphasised that the US is putting a great deal into relevant technology and that many states, cities and companies are working hard to reduce emissions, even though the US did not sign the Kyoto Protocol, uses gas-guzzling cars and has one of the highest per capita emissions of any major country. This sharing of information is not, of course, a task for just one individual, and I use my own experience simply as an illustration. There will be a major task of building impartial scrutiny from internationally respected institutions, so that each country can understand the effort that others are making.

Some practical politics

The most significant countries or regions in terms of population and emissions are China, the US, Europe and India. (I take Europe as a political entity for the purposes of thinking about key movers in an agreement, and although India's current emissions are below those in Brazil, Indonesia and Russia, they are growing rapidly and the population is very large.) A mutual understanding between the two largest emitters, the US and China, will be absolutely fundamental to any agreement. Given its wealth, its technology, its historical responsibility for emissions and that per capita they are now double that of Europe, four times that of China and more than ten times that of India, the world will look for leadership to the US. China in particular, crucial though it must be to any deal, will understandably look to the US not only to commit itself but also to *demonstrate* its commitments, before joining a global deal.

What are the prospects for action in the US, and what are the obstacles? First, it is important to recognise that the US is a very varied entity. There are many states committed to strong action to reducing emissions: twenty-eight have enforced climate action plans – California being especially prominent – and nine others have emissions targets. The mayoral leadership in a number of cities such as Las Vegas and New Orleans has also given clear commitments to disclose or reduce emissions. Many US firms have made strong progress in energy efficiency and are leading the way in technology for low-carbon energy.

It is true, however, that the Bush administration has slowed progress by contributing to the hesitancy of the major countries of the developing world about a global deal. At the Japan summit of the G8 in July 2008, the G5 (China, India, Brazil, Mexico and South Africa, who now regularly attend part of the G8 discussions) refused to join the commitment to halving global emissions by 2050 unless they saw the specific targets and action that the rich countries were ready to take; progress in the rich world taking on the necessary targets was impeded primarily by the US. At the UNFCCC conference in Bali in December 2007, it looked as if, right up to the last minute, the US might block the launch of negotiations for a successor to the Kyoto Protocol, and the setting of the target for agreement in December 2009 in Copenhagen. Kevin Conrad, Papua New Guinea's ambassador for climate change, famously said to the US delegation: 'If you cannot lead, leave it to the rest of us. Get out of the way.'

However, during 2007 there was also some movement and the US launched the Major Economies Meeting on Energy Security and Climate Change, designed to move discussion forward among the major emitters. The Bush administration was also the initiator, together with Australia, of the Asia–Pacific Partnership on Clean Development and Climate.

It is striking that an administration that had been so difficult for so long on this issue showed real signs of movement in its last couple of years, even though it was rather late. In my view this change is likely to have arisen from a combination of four factors. The first is the overwhelming nature of the evidence on the causes of climate change and the gravity of the risks. The second is that pressure in US politics was building and that to maintain either denial or obstructiveness, or both, would be politically damaging. The third was a growing understanding of the risks and volatility in markets for hydrocarbons, with an increasing political focus on energy security. The fourth is pressure from large parts of US industry for clarity of policy and thus the encouragement to pursue new low-carbon investment opportunities. The creativity and vigour of those involved in the new technologies across the United States is remarkable. These reasons give grounds for optimism on the direction of future progress. There may have been a fifth reason, in that the administration realised this issue was damaging the country's position and reputation in the international community at a time when it was already in difficulty.

For any US administration, overcoming the misconception that policies to reduce emissions will harm the economy, and making convincing arguments that domestic support for change is possible without resorting to trade barriers, is a key challenge. Competitiveness issues affect only a few sectors to any serious degree, although the political voice of these sectors is relatively loud. The voice of potential winners in the private sector, on the other hand, is not being heard loudly enough. A powerful case on the new opportunities for America from strong action on climate change, and addressed directly to his fellow Americans, has been made by Tom Friedman in his book *Hot, Flat and Crowded*, which in autumn 2008 was number one on the non-fiction best-seller lists in the US.

The world saw the US fail to ratify the Kyoto Protocol after the US delegation agreed to it in 1997, because the administration could not get it through Congress. The picture in Congress has been transformed in the intervening period, and many members of both the Senate and the House show deep understanding of both the risks and opportunities and a real commitment to lead. And, of course, from January 2009, there is a new president. But, after seeing what happened in the US on the Kyoto Protocol following 1997, other countries considering their own position will reflect very carefully on what the US, taking the administration and Congress together, can actually deliver in relation to commitments made by US representatives at international summits. It will, therefore, be critical for the new administration and Congress to move quickly and it must be made clear where the US is headed. This could unlock potential international commitments, particularly from developing countries, in the near future. Such progress could also allow the US and China to come to a common understanding soon, and lay the foundations of a framework for action. This is a moderately optimistic picture, but it is far from impossible.

What of China? Since the mid 1990s, beginning with the premiership of Zhu Rongji, it has recognised the very serious problems presented by the environment.[1] China has seen, for example, extensive soil erosion on the Loess plateau, air pollution in cities, the pollution of its rivers, and major problems of water management and supply across the country. Concern about water has been a recurring issue in China's history, and these worries are increasing as patterns of rainfall and river flows alter. China is vulnerable to droughts, floods and rising sea levels, and there is little doubt that it understands the dangers.

China is reforesting not deforesting; it has a 20% energy-to-output reduction target for its eleventh five-year plan (2006–10) and is considering pathways for low-carbon growth in the context of the twelfth five-year plan; at the end of 2006 it introduced a tax on the export of energy-intensive goods, equivalent to a carbon tax of around $40 per tonne of CO_2.[2] There are emissions regulations on cars which are sufficiently strict to preclude the import of most US cars. The Chinese government has pursued a consolidation of the steel sector, with state-owned mills acquiring smaller independent producers, which allows for efficiency standards to be implemented more directly in the smaller production facilities. While the leadership takes the view that growth must continue for the stability of the economy, and for raising the standard of living in a nation that is still very poor, with hundreds of millions living on less than $2 a day, the eleventh five-year plan states very clearly that 'harmony' between the environment and growth is crucial. In June 2007, an action plan on climate change pulled together many of the elements noted above into a strategic vision for change.

China also knows that there will be no global deal unless it plays a very strong role: it is a potential deal-breaker. If you know you are vulnerable if a deal fails, and if you know you are a potential deal-breaker, then you have to concentrate. China understands this clearly. And now so do all the major players as well, particularly the US, Europe and India.

Nevertheless, there is of course the real problem of demand for energy in a rapidly expanding, industrialising economy, and the costs of action as a percentage of GDP are likely to be higher in China than for many other countries. Between now and mid-century the Chinese economy is likely to grow by a factor of at least ten. (At 7% per annum growth, an economy will double every ten years and China has been growing faster than that over the last three decades: four decades of doubling each decade would give a factor of sixteen; thus a factor of ten may in fact be moderately conservative as an estimate of expansion between now and 2050.) China will have to divide its overall emissions by a factor of around two if it is to get into the region of 2 tonnes per capita by 2050, given that its emissions are now over 5 tonnes per capita (and allowing for fairly modest population growth of around 20% over four decades – China's current annual population growth is around 0.5%). If total emissions are to halve, and the economy will grow by a factor of at least ten, then emissions per unit of output will have to fall by a factor of twenty. That means a 95% reduction of emissions per unit of output.

This is a very radical change, essentially decarbonising the economy. Preliminary estimates from the IMF suggest that costs might be between 1.6 and 4.8% of real GNP by 2040.[3] Change of this magnitude will require collaboration and support from the rich countries, in particular through technology and the financial flows that could come from carbon trading. China is examining the costs of change carefully, and has a strong focus on the technological changes that will be necessary for low-carbon growth, and therefore vital to a global deal. It is significant here that whereas in many countries politicians have, for example, a legal or business background, in China the leadership contains many engineers, for whom technological issues have a very high profile. Given that the US is the most technologically advanced major country in the world and many of its enterprises are moving strongly in this area, collaboration over technology will be one of the key elements in a US–China understanding and for a deal to be reached.

The EU and many of its constituent countries have, for the most part, been in the vanguard on action to tackle climate change. The council of heads of governments made a crucial commitment in March 2007 of 20% reductions in emissions relative to 1990 by 2020, and 30% in the context of strong movement by other countries – essentially as part of a global deal. Specific measures were set out in January 2008, including greater industry coverage and expanded within EU trading possibilities for the EU Emissions Trading Scheme, targets for industries outside the EU ETS, and legally enforceable rules for renewables.[4] At the G8 Summit in Heiligendamm, under the chair of Germany's Chancellor Angela Merkel, there was a statement on the global goal of a 50% cut by 2050 (although the base date was left open). The commitment to those cuts was strengthened at Japan's Hokkaido G8 Summit in July 2008. France has had its *facteur quatre* (dividing by four between 1990 and 2050) since 2004. In October 2008, the UK raised its commitment from 60% cuts by 2050 to 80%.

We can expect the EU and its member countries to continue to drive forward action on climate change. There will, of course, be political difficulties and obstacles to be surmounted and serious challenges of delivery (some emerged in December 2008 and were overcome), but the course in Europe seems to be well set and many interesting instances of practical progress exist, whether they be onshore wind for electricity generation in Germany and Denmark, or tighter regulations on car emissions which are on the way. The EU is committed to increasing the share of renewable energy to 20% with a minimum target of 10% biofuels by 2020. It also committed itself to achieving a 20% reduction in energy consumption by 2020. Under the Action Plan for Energy Efficiency (2007–12) it estimates energy savings of 27% in residential buildings, 30% in commercial buildings, 25% in the manufacturing industry and 26% from transport.[5]

In India, as in China, there is a growing understanding of the vulnerabilities that the country faces. The main rivers of north India, and indeed of China, Bangladesh and Pakistan, rise in a few hundred square kilometres of the Himalayas and the glaciers are already retreating rapidly, by 15% in the last forty years.[6] This part of the Himalayas has been described by Prime Minister Manmohan Singh as the water tower of Asia. If its capacity to hold water continues to reduce rapidly, India and its neighbours with rivers also rising in that area face unmanageable torrents in the rainy season and dry rivers at other times. The monsoon is likely to be seriously disrupted, and coastal cities are vulnerable to more frequent severe storms and to sea-level rise. India's future development depends crucially on a global deal.

While India's emissions are only around a third of China's, the country's size and growth ambitions mean that it too will be critical to the establishment of the global deal, and is also a potential deal-breaker. Thus, minds in India are increasingly concentrated on this issue. The prime minister established a cabinet climate change committee in May 2007, and an action plan was published in June 2008 where the Indian government pledged to devote more attention to renewable energy, water conservation and preserving natural resources. Climate change was an important theme of both the finance minister's budget speech of 29 February 2008 and later of the prime minister's Independence Day speech on 15 August, a significant event in India's political calendar, when Manmohan Singh asked for a national consensus to be developed to tackle the issue.[7]

India still has around a third of the billion people in the world living on less than $1 a day, and is now showing a strong and sustained capacity for growth which is making inroads into that poverty, and this should be warmly welcomed by all. Thus, India's growing interest in climate change is generally qualified in international forums by a statement along the lines that it will do what is in its self-interest and that development comes first. But increasingly that is not the end of the discussion, and India is engaging on the structure of the global deal.

Some of those in India, and not only in India, who are on the left portray climate change as largely an upper-middle-class issue for the future, whereas poor people, they argue, are concerned about poverty in the here and now. Some of those on the right see dangers of a command-and-control dirigiste approach re-emerging in a green uniform. Both perspectives are thoroughly misguided: it is poor people who will suffer most and markets which fail should be fixed, not abandoned. Low-carbon growth is the *only* growth strategy for the medium and long term. But these arguments, made strongly throughout this book, must be brought into the broad political domain if action is to be sustained and the challenge of linking growth, mitigation and adaptation is to be met.

The politics of climate change in many other important countries is changing rapidly. In June 2008, Brazil launched its own plan to halt deforestation and sought international support. Mexico, too, has been taking a lead in Latin America through its own detailed country study of action on climate change. Guyana is seeking international support to manage its forests, and Costa Rica has pledged 100% reduction in emissions by 2021.

I could go on with examples from many other areas of the world, but the essential point is that public understanding and political discussion, both of the phenomena and the policy issues involved, has moved on remarkably in the last three or four years. So too has the pace of development in technology and associated investments. This book is about what can and should be done, and how obstacles can be overcome. To that end I have just given a fairly positive assessment of how the politics are moving now, and can move in the future. If we focus too heavily on the negatives and say 'It is all too difficult', 'People are too short-sighted' or 'Narrow self-interests will always prevent action', then pessimism will become a self-fulfilling prophecy. At the same time, however, if the very substantial obstacles are indeed to be overcome, then the negatives have to be recognised, carefully understood and analysed, and measures to overcome them proposed. As the details of a global deal are negotiated, flexibility, imagination and creativity, combined with an internationalist and collaborative approach can find a way through to a conclusion. I think the broad description should be along the lines I have described but there are many possible configurations.

Arguments and analyses

Country by country, the public and the leadership have to be convinced that low-carbon growth is possible and attractive, and that the adjustment from current patterns can be managed. The case for both must be made not only analytically, but also by demonstration. At a New York gathering of the ECOSOC (the UN's main economic and financial committee) in June 2008, one of the representatives from a large developing country put it to me very directly: we have the rich-country high-carbon growth model of the past, he said, but we do not have the low-carbon growth model of the future. It is now a matter of great urgency to provide an analysis of what low-carbon growth looks like for each country. That is essentially a task for the nation itself, but bilateral aid agencies and multilateral institutions can and should provide major support for developing countries, and help to share strategies and options for all countries.

A second key task will be to prepare analyses of the implications of varying forms of agreement for different participants ahead of critical forthcoming summits, especially Copenhagen in December 2009. International negotiations on trade have a long history and it is normal practice for countries to prepare in this way, but in some respects consequences relating to trade are more easy, or at least less difficult, to predict, compared to the economic effects of both action and inaction on climate change. This is partly because trade analysis and modelling has been taking place for a long time, and because of the conceptual simplicity of the models and phenomenon, at least relative to climate change. As the crunch comes in climate change negotiations and specific agreements have to be made, countries will want to understand what different versions of the specifics mean for them. This also must be a high priority for the analytical community: a very positive example here is Project Catalyst, initiated by the Hewlett Foundation (in collaboration with the consultants McKinsey) to work on details both of low-carbon growth and the possible structures for an agreement. We shall need as much of this kind of analysis as possible at the country level, too.

Looking beyond the next year, the power of the example will be of great importance, which is why it is right to emphasise the wisdom of conditionality imposed by developing countries on rich nations relating to strong targets, demonstration of low-carbon growth, carbon finance and technological sharing. But over the period up to December 2009, the rich world must be really convincing on its intentions and as specific as possible on measures and commitments, if the political breakthrough in the developing world is to come.

While the two analytical priorities are low-carbon growth and the implications of different agreement structures, there are a number of other arguments where careful marshalling of evidence can help deal with opposition and confusion.

One, which is mercifully diminishing but still has to be firmly dealt with, is the claim that climate change is not a problem. We should be aware that as the time to make a deal nears, somebody somewhere will re-emerge to say, 'There is no real danger, let us not bother with this.' The counterarguments to this nonsense are ready and clear, but they will have to be used again.

A second area is the recurring issue of *competitiveness*, whereby it is argued that if one country acts earlier than another, its industries and firms will be at a disadvantage when competing internationally. For the most part such claims are slogans without numbers, and usually grossly exaggerated: if we analyse the figures, the extra costs for the vast majority of industries are likely to be in the order of 1–2% of global GDP per annum over the period to 2050. That would be similar to a one-off 1–2% increase in costs of production, the type of change which can be accommodated. Indeed, sustained changes in exchange rates which firms do cope with often have larger cost effects than this, and the magnitude is small relative to wage-rate differentials between, say, Europe and China which are a factor of five or ten.

For just a few sectors such as aluminium, steel, cement and one or two others, costs are likely to be higher and transition arrangements would have to be carefully managed. There are a number of options. Rich countries could, for example, propose a particular international deal for those industries, with three elements: a slower move towards auctioning of allowances under cap-and-trade schemes in their

countries; collaboration with firms in developing countries to introduce new technologies; eventual introduction of countervailing tariffs
in, say, 2020 if action to reduce emissions in developing countries are
not implemented. This package could give firms time to adjust and to
install the new equipment which will have lower emissions and thus
incur less carbon cost while maintaining profitability along the way.
This is the type of change that policy is intended to generate. Firms that
are liquidity-constrained and have existing capital stock cannot simply
leap to new technologies. Alternatively, the industries themselves
could be encouraged with some support to come up with their own
plans for speedy reductions in emissions worldwide. The competitiveness issue can be dealt with, in the few cases necessary, by particular
sectoral, time-limited policy packages. The fact that a small number of
industries are particularly carbon- and energy-intensive, for whom
costs of adjustment may be more substantial, cannot be a reason to
stop progress towards a global deal.

Some commentators argue that there will be a massive relocation
of industry to countries with weaker climate change policies. There
are many factors influencing firms' location decisions (investment
climate and bureaucracy, financial and tax systems, legal and administrative systems, transport costs, access to markets, access to raw
materials, access to skilled labour and so on) and the carbon regime
would be just one factor. Further, the empirical evidence on this type
of mobility from careful studies of the data indicate that such effects
are small.[8]

Others will raise difficulties about strong reliance on carbon trading,
arguing that there will be corruption over permit allocations, and
reductions will be fake. And one often hears language to the effect
that 'this is a get-out-of-jail-free card for rich countries'. Another is that
traders will 'rip off' poor countries with low costs of abatement and
capture the returns. Again, as with competitiveness, the answer is to
deal with the genuine issues and refute bad arguments and not to
abandon an attempt to use markets with the great advantage they can
bring of lowering costs of action.

There will be important challenges for market regulation just as
there are, for example, in insurance and financial markets. The other
major price instrument in this area is carbon taxes. They have their
role to play but they too have their problems: finding international

coordination on taxes is never easy; taxes do not give the same degree of confidence in quantity reductions; national governments may not be convincing in their continuity over time, given that they may become a domestic political football; and taxes may not guarantee a flow of finances from rich to poor countries. The challenge is to find the right combination of taxes and trading and then to ensure that both elements are well managed. Monitoring and regulating the carbon market will be an international challenge, but one we must rise to; open and transparent auctioning of permits can reduce corruption. None of the instruments will be perfect, but in their design and management there is much we can do to make them work well and, hence, keep down the cost of the transition to a low-carbon economy. There are real choices and options, and sound analysis and creativity should be able to find solutions – we should not be overly rigid in insisting on particular mechanisms.

There will be those who argue that policies will inevitably be badly chosen and implemented, so raising costs of action. There is indeed a danger and the response of policy analysts and economists should be to provide clear and strong guidance. That is what many are doing. Edenhofer (see Chapter 6) has shown that costs of action are much higher if we exclude CCS for coal, and the Eliasch Report suggests overall costs of holding down concentrations may double if deforestation continues at its current pace. Peace and Weyant, for example, have provided careful and important analyses of how costs rise if sectors, countries, technologies, or gases, are excluded.[9]

In the arguments and analyses building towards the deal there is much work to do. Objections cannot be brushed aside, but must be dealt with seriously with careful analytic work and policy measures where relevant. This does not mean buckling to the special pleading of self-interest or accepting arguments that are wrong. And it does not mean we have time to spare – December 2009 is very close, and the problem itself is urgent.

Making policy in a turbulent world economy

Policy on climate change is for now, and for the medium and long term. To provide a clear framework for investors and consumers to take measured decisions, there must be stability of policy and a clear sense of direction over the next few decades. How will forming and sustaining policy, and the changes and restructuring required, be affected in a period when confidence in the world economy has been shaken, many parts of the world face recession and prices of hydrocarbons are likely to be high over the medium and long term? There are indeed difficulties in making long-term policy in such circumstances, but also opportunities.

First, and this will require clear and strong political leadership, we must draw two key lessons from the financial crisis itself. The first is that the crisis has been developing over twenty years and surely tells us that ignoring risk or postponing action is likely to store up trouble for the future. The implications of unmanaged climate change will appear longer in the future than twenty years but action in the next few years is crucial if the more extreme risks are to be avoided. The impacts of unmitigated climate change will be much larger than those of the current financial crisis.

Second, we shall have to grow out of a recession which may be extended: this is years, not months. We will need a driver of growth which is genuinely productive and valuable and which can generate a further period of expansion. The clearest example is surely low-carbon growth. It would be very different from the dotcom boom of the late 1990s and the selling of half-baked websites at inflated prices, and the housing bubble of the last few years, conflated with the selling of financial instruments that few, if any, understood. With the right government policies, those discussed in this book, the private sector stands ready with a whole wave of potential investments and new technologies. The emphasis on sound policy on climate change must not be reduced because some elements are, quite rightly, being integrated with a package for economic recovery – for example, mistakes such as the subsidising of US corn ethanol, or the protectionism against imported biofuels, should not be sustained or repeated.

There are other issues too raised by recent experience, particularly concerning hydrocarbon prices and energy security. When times are hard or hydrocarbon prices are high, energy efficiency and economising on hydrocarbons will of course be still more attractive, and the relative profitability of low-carbon alternatives such as wind, solar and nuclear will increase. These effects of higher hydrocarbon prices, tough though they are for many people and countries, are helpful for making policy on climate change. We should not be deluded by short-term price falls of oil in the slowdown of 2008–9 that prices will stay low (see, e.g. the World Energy Outlook of the International Energy Agency, 2008).

On the other hand, there are key aspects of incentives which become more difficult. With high oil prices, the exploitation of resources such as oil sands and oil shale – which consume a great deal of energy to extract, and emit between 20–200% more greenhouse gases in the process of refining into gasoline or diesel – starts to look attractive.[10] Changing relative prices within hydrocarbons is problematic too – while coal prices have risen in 2007–8, they have not risen as much as oil and gas, giving an incentive to switch to coal, which is around twice as polluting as gas (in terms of greenhouse gases) in electricity production. This makes it all the more urgent to develop the technology, logistics and legal framework for CCS for coal very quickly, and provide the incentive structures through carbon pricing to use the technology.

The political premium on energy security will rise in turbulent times, and from this perspective renewables become more attractive – but the same applies to coal for the many countries which have it. Hence again the importance of CCS. Energy security and responsibility on climate change thus pull in the same direction, provided we are successful with CCS.

It will be politically more difficult in some respects to add further to energy or electricity costs by pushing through a price for greenhouse gases. On the other hand, a price of $40 per tonne of CO_2 translates to around 40 cents on a US gallon of petrol, which in percentage terms is only 10% of the summer 2008 price of $4 per gallon in the US, compared with 20% relative to the $2 price a year or two before. Roughly speaking, the same $40 tax per tonne of CO_2e translates into $20 per barrel of oil (when the full damage from all gases, not just CO_2, is reflected in the cost), so the increases in oil prices from around $60 per barrel in mid 2007 to around $140 in mid 2008 are much higher than those which would follow from carbon taxes. Price increases as a result of carbon pricing should not therefore be seen as so large as to be politically unacceptable – at least, wise political communication and leadership should be able to present the argument convincingly. When oil prices are high it may not be politically opportune to introduce a carbon tax, but it may be a good time, for example, in 2009 to announce that as they fall, then that fall may be partially offset by such a tax.

It is important to be clear that a rise in oil, gas and coal prices does not remove the argument for a carbon tax, or for trading. Whatever the price of hydrocarbons, their use damages others unless the carbon is captured, and the argument for a price for carbon therefore remains valid. For example, using coal with CCS will always be more costly than using coal without CCS, hence a price on carbon is necessary to give an incentive to use CCS.

Looking beyond the current difficulties in world and oil markets, we must remind ourselves that over the next forty years, when policy on climate change must be steady and strong, there are likely to be a number of ups and downs in the world economy. After the first oil crisis in 1973, for example, the share of GDP spent in the US on energy almost doubled, contributing to throwing the US economy, and the rest of the world, into recession. The very fact that the rising oil prices of 2008 contributed to our current difficulties reminds us that establishing a long-term trend away from hydrocarbons will make us less vulnerable to the vagaries of those markets in the future.

The financial crisis of the last year or so and the likelihood of high oil prices over the medium term should make us ever more committed to the movement away from hydrocarbons and towards low-carbon growth. Leadership requires strong focus on a clear sense of direction as the inevitable cycles occur. To go flaky on policy would undermine the investment that is essential for this growth.

The politics of sustaining agreement

Building an agreement is the first and most urgent part. It must also be sustained. Part of a deal's ability to be sustained must lie in the agreement itself and its incentive structures; partly from concern that action by one country to undermine the agreement can imply subsequent hardship on a global scale; partly from political pressures; and partly from institutional structures.

Assuming the time horizons are sufficiently long, the agreement itself should generate gains for most or all countries. This is not a zero-sum game where a gain to one is automatically a loss to another. The potential gains are very large. A market imperfection, if corrected, overcomes inefficiency – policy can make everyone better off. Correcting a huge imperfection such as this carries huge gains. Although the benefits to a country that behaves in a way that sustains the global deal will be enjoyed largely by the future citizens of that country, and all others, current generations may, and perhaps should, see gains to future generations as something that makes themselves better off. If we wish, we could see the gains to future generations accruing from a global deal that we have helped create and sustain as justifying a little less saving of the more conventional kind.

In the shorter term the incentive structures should be such that even over modest time horizons, poorer groups see a direct incentive to participate. That is the point of carbon markets and sharing of technologies. There will be some shorter-term costs, and for equity they should be focused on the rich countries. Even here, costs can be spread between generations and they should derive some satisfaction in the knowledge that as a result of their responsible behaviour their successors will have a better climate.

Individual countries participating in collective action might also see gains, via an enhanced reputation for responsibility, in their international interactions on other dimensions. In other words, if a country fails to cooperate or deliver to the extent it has promised, it might be less trusted in other discussions and agreements on, for example, trade, investment or financial regulation. It would probably be a mistake to make this formal, as it could lead to major complications both in climate change and other areas. For example, international law on trade disputes have resolution mechanisms, and burdening them with formal climate change issues might make decisions and action hopelessly complicated. Nevertheless, international collaboration and diplomacy has commonly had an element of these broad and implicit trade-offs. It is indeed one way in which decision-making in the EU functions – reputations or lack of willingness to find a shared view in one area have consequences for another. This perspective is another reason why it is important for heads of governments and finance ministers to be directly involved – a minister in charge of a particular set of responsibilities would not necessarily be empowered to make these trade-offs. Whether or not trade-offs are implicit or explicit, where countries interact over many issues and many time periods, reputation matters.

Over the medium term, I do not think that a formal structure which is empowered to administer 'international punishment' or behave as an 'international enforcer' is likely to be either feasible or desirable. It would sour relationships and risk withdrawal by a number of participants. The challenge is to find ways of promoting, encouraging and incentivising a collegial approach.

The second of the sustaining forces, the worry about undermining a global deal and the overall consequences of so doing, illustrates an approach to a 'game-theoretic' perspective on these negotiations over the long term. The simplest form of game theory in this context sees participants in the international game focusing only on their narrow self-interests and making very simple assumptions about the behaviour of others, that is, if I cheat then others will continue as before. In that narrow approach, the 'free-rider' problem looms large. In other words, one player assumes that once an agreement is established he can cheat and save money by not cutting emissions as promised, that the others will be more responsible and maintain their efforts, and thus he will not suffer very much from the extra emissions he has made. There are three things wrong with this simple version of the free-rider story. First, many or most players are likely to worry that if they cheat then others may withdraw from the agreement, and the consequence of their 'minor' increase in emissions will turn out to be major and very damaging increases globally. Second, as agreements and action start to build and new markets are created, the players seen to be irresponsible risk being excluded from key opportunities and major areas of future growth. An attempt to free-ride may prove costly for those who try. Third, many participants will look beyond their own interests and worry about the consequences for all. This is not to diminish the importance of the free-rider problem – it requires thought, design and watchfulness to be held in check – but even the narrow version of the argument has theoretical limitations. Indeed, the reputational effects and implicit sanctions outside climate change agreements can play their role in a game-theory discussion.

The third sustaining force is domestic political pressure. The issue was raised as a force for building an agreement, and it can also sustain one.

Many people in many countries understand the meaning of responsible behaviour on climate change, and demand that their politicians and decision-makers follow standards of responsibility. They are prepared to challenge or vote them out of office if they fail to meet these standards, and to support them if they act responsibly; the recent rejection of John Howard as prime minister of Australia and the re-election of Arnold Schwarzenegger as governor of California are examples. People can bring pressure on climate change, and politics can respond to that pressure. This will be a real force in sustaining agreement.

International institutional structures

If John Maynard Keynes and Harry Dexter White were sitting down to design the key international institutions now, instead of in Bretton Woods in 1946, they might end up with a different three. Instead of the IMF, the World Bank and the World Trade Organisation (WTO) they would now, I think, have one institution combining the World Bank and IMF, the second would be the WTO and the third a World Environment Organisation. We are not, of course, in a position to start with a 'clean sheet' on the institutional front, but we must ask ourselves whether the task of taking forward an agreement on climate change, in the context of other environmental challenges facing the world, requires a new organisation. However, the building of an agreement is so urgent it will have to be carried out with existing institutional structures. Institutions take time to be agreed and built and to deliver action, and we require an agreement by the end of December 2009: we must work for Copenhagen within the UNFCCC.

Looking forward, a new institution should be defined and created on the basis of its functions, and after asking the question as to whether those necessary functions cannot be fulfilled by existing institutions, modified as necessary. I have spent ten years of my working life in two very good international institutions, first as chief economist of the European Bank for Reconstruction and Development, founded in 1991, after the fall of the Berlin Wall, and then as chief economist of the World Bank which goes back to the 1940s and the aftermath of the Second World War. I participated in the building of the first, particularly on the strategy front, and helped set a revised course for the second, a long-established institution. In both positions I had intensive interaction with the UN agencies, the EU and many national governments of both developing and developed countries. Notwithstanding my own two positive experiences, I hesitate to recommend a new institution given that the setting up and running costs are high, but nevertheless I think, on the basis of the required functions, it will indeed be necessary in the next five or ten years. A global agreement along the lines described and, indeed, most if not all the plausible agreements one might imagine, would need to take on the following tasks:[11]

- the development of global and distributed emissions targets, timetables and milestones
- the detailed design of a new, greater scale and more extensive CDM mechanism
- the establishment of the trading element of any cap-and-trade system
- the creation of systems to supervise, monitor and verify delivery against commitments including any pilot global sector agreements
- the emerging forestry carbon regime
- the coordination and increased funding of advances in climate change-related science and low-carbon technology development
- the development of an improved and joined-up understanding of the potential local risks from climate change and the responses as they develop
- the development of processes for dispute resolution
- the examination and proposal of coherent links concerning other global challenges, including development, trade, biodiversity, food and energy – relations with other international institutions will be important here
- the analysis and proposal of ideas for the funding of the range of activities described and for rules of operation of the allocation of funds, which will change over time as lessons are learned and circumstances change

It would be possible in principle to run some of these activities through other international institutions, but there are two problems with doing that. First, coherence within climate change activities is of great importance, and without a clear primary responsibility in one place it would be very elusive. Second, there would be a real danger that its priority would slip relative to other shorter-term issues in these other organisations.

As with many international bodies, its governance, instruments for action, focus on results, links with other organisations and account-ability will be of great importance. There is much to learn from the experience of other institutions. The design of a new one must integrate well with, and absorb, some of the functions of existing institutions, if overlaps and wasteful turf battles are to be avoided. The institutional design will be a major challenge and one where detail matters. They will have to be sufficiently flexible to be able to respond to strong learning processes and to a changing environment, while retaining credible long-term frameworks.

I would emphasise just one issue here, and that is governance. The world of Keynes and White in 1946 was dominated by a few rich countries and by the victors of the Second World War. Much of the world was colonised. The economic, population and social structures of the twenty-first century are very different. Climate change is a subject covering the entire planet and 8 billion of the 9 billion who will be on this planet in 2050 will be from currently developing countries. If this new institution is to work effectively, it must start with a governance structure and with rights to shape and determine decisions which reflect the whole world in an even-handed way.

If a new organisation is designed with this in mind, it is likely to make a contribution way beyond the environment and climate change: it could revitalise the whole international institutional structure, not only by giving an example of how to be effective, accountable and well governed, but also because it can show that if we act together we can take on the most difficult of problems, for the good of the world as a whole.

CHAPTER 10

A planet in peril

In his speech in Chicago on the night of his momentous election victory on 4 November 2008, Barack Obama provided a sober assessment of the tasks ahead: 'The challenges that tomorrow will bring are the greatest of our lifetime – two wars, a planet in peril, the worst financial crisis in a century.' I have described how we can rise to one of these challenges, offering a 'blueprint for a safer planet'. If we follow the route I have tried to chart, or something similar, as I believe we can, we will not only protect the planet for our grandchildren, we will also reduce dramatically the severe threat of global conflict that unmanaged climate change will eventually bring.

The cooperation across nations that this path demands can itself herald a new era of internationalism that could make wars less likely in our generation, and also enable us to tackle more effectively the other problems that require a global response, including trade, financial stability, disease, drugs, nuclear proliferation and poverty. If we start now in a strong and measured way to create the activities, technologies and investments that are necessary to build a low-carbon economy, we can give the boost to demand that the world economic slowdown requires, and lay the foundations for the key driver of sustained growth over the coming decades – new environmentally friendly technologies.

At the end of his historic speech, Obama took the long view and, referring to a famous supporter aged 106, asked, 'If my daughters should be so lucky to live as long as Ann Nixon Cooper, what change will they see, what progress will we have made?' If we do not set ourselves strong targets now, and institute the policies and the global deal necessary to move towards them in a clear, purposive and measured way, the answer to this question will be a degraded planet, a hostile environment, and a world of conflict and growing poverty. On the other hand, we can achieve, by such action, a safer, cleaner, quieter, more biodiverse world with growing incomes; a world where we can win the other major battle of our century, the fight against world poverty. We either succeed on both, or we fail on both; there is nothing that is stable in between.

These are our choices. What are the chances of success? We can see our way forward in terms of the technologies and economic policies. We are learning quickly and will learn much more along the way, but the basic elements of the road ahead are clear. The key obstacle is the political will: do we, together, have the wisdom and collaborative spirit to take, and deliver on, the necessary decisions? I do not know the answer, but I can see the likely consequences of pessimism and of optimism. First, if we assume people and politicians will be irretrievably short-sighted, quarrelsome and narrow in their judgement of their interests and act accordingly, then our pessimism will be self-fulfilling. Second, over the last two years or so there have been increasing grounds for optimism in the growing understanding and commitments of countries around the world; if we recognise the progress that others are making and act in mutual support, we have a good chance of responding on the scale the planet requires. We will find out in the coming months and years which of these forces are dominant and whether the latter will be fast and strong enough to take us to an effective, efficient and equitable global deal in Copenhagen in December 2009. These are not forces over which we have no control: it will be for us all, working together, to determine the answer to the questions that Barack Obama put so clearly as he looked to the future.

How are the countries of the world now seeing the challenge and their approach to a global deal? The change in understanding on climate change in the last few years has been profound. I have been watching closely and participating actively in these discussions in most of the key countries. At the G8 Summit chaired by the UK in Gleneagles in July 2005, part of which I attended, the two key subjects were support for Africa and managing climate change. The engagement and understanding of many of the world's leaders on climate change at that time was perfunctory. There was particularly strong interest and activism from the UK and France, a wish to move forward from Germany and Japan, but little real enthusiasm from the USA, Canada, Italy and Russia. Indeed, some of the leaders looked bored.

Since then the involvement of Germany and Japan has intensified strongly with Angela Merkel, the chancellor of Germany, chairing the G8 Summit in June 2007 in Heiligendamm to deliver the statement of a goal of 50% cuts in emissions by 2050, and with Japan leading the strengthening of that commitment at the G8 Summit in Hokkaido in the summer of 2008. In its Council of Ministers in spring 2007, Europe made its commitment to 20% reductions in emissions by 2020, relative to 1990, with 30% in the context of a global deal.

The most important countries in shaping a global deal will, however, be the US and China, the world's two largest emitters. The election of Barack Obama, as his campaign and victory speech made clear, will transform the position and also, importantly, the perceived position of the United States. The world can now believe that the leadership of the US understands the challenge and its urgency, the crucial role of its leadership and the importance of internationalism. The decision-makers in China – and I interact regularly and strongly with those working on the issues – are keenly aware of the potential effects of climate change on their future and on their pivotal role in reaching an international agreement. China is an intensely practical and analytical country in its decision-making, with many engineers among the leadership. They are now examining carefully how they can cut emissions, together with the costs and technologies and the type of support they need. When I was in Beijing in October 2008 at the same time as the Danish prime minister, the Chinese leadership indicated their strong commitment to reaching a constructive deal in Copenhagen in December 2009.

If the US makes strong progress during 2009 towards a cap and-trade bill, embodying the 80% reductions by 2050, relative to 1990, that Obama has indicated is necessary, then the rest of the world will have greatly increased confidence in US commitment. If, at the same time, China begins in 2009 the construction of its twelfth five-year plan (2011–15), with a strong theme of low-carbon growth, then others will see that it intends to intensify its actions on climate change, building on the foundations of the strong energy-efficiency targets of the eleventh plan and the climate action plan published in June 2007. There is now a real possibility that an effective and mutually supportive understanding between the US and China could grow vigorously during 2009. That will be critical to the prospects for global agreement.

I do not want to pretend that agreement will be easy; on the contrary, the road from here to Copenhagen and beyond will be very tough and full of obstacles in the form of mutual resentment, particularly towards rich countries for their historical responsibility for high-carbon growth, and in narrow perspectives of self-interest.

Central among these obstacles will be the argument that the first priority is to deal with the current economic crisis and that action on climate change can be postponed. Often this argument comes from those who are, in any case, not keen on taking action and who use the economic crisis as an excuse. But the argument is wrong and it must be confronted. There is no doubt that the economic crisis is of great seriousness and requires coordinated and strong action both nationally and internationally; the error lies in seeing action on the economic crisis and on climate change as conflicting. They are not, and the economic crisis becomes an obstacle to urgent action in climate change only if we let it be so by failing to put the arguments clearly.

There are two key lessons we should learn concerning the relationship between the economic crisis and the planetary crisis. The origins of the financial and economic crisis the world now faces go back twenty years. Financial markets were deregulated; complex derivatives were erected on the back of a housing bubble; and ultimate lenders were separated from ultimate borrowers; the spending on the basis of the bubble was facilitated over the last decade by a willingness from the rest of the world to finance the growing deficits of the US, a willingness which in East Asia was driven in large measure by a desire for substantial reserves which arose from the experience of the East Asian crisis in the late 1990s. The lesson should surely be that the longer risk is misunderstood and ignored, the bigger the consequences when the crash occurs. We must not make the same mistake with climate change.

The second lesson from the crisis is that we shall have to find a driver of growth to take us forward in a sustainable way. A large short-term fiscal boost, coordinated around the world, is essential but will not be sufficient to set us on a new growth path. The new technologies and investment opportunities of low-carbon growth will be the main drivers of sustainable growth in the coming few decades. These investments will play the role of the railways, electricity, the motor car and information technology in earlier periods of economic history. Let us organise our emergence from the crisis by investing in the shorter-term projects, such as energy efficiency, which can generate demand

and employment quickly and by bringing forward some of the energy and transport infrastructure investments which can lay the foundations for medium- and long-term growth. While there is no doubt that a major consumption boost is necessary as we tackle the severe problems of insufficient demand in the very short run, we must look forward to the investments which can sustain real growth in the future.

We can and must, now and simultaneously, handle the short-term crisis, foster sound growth of the economy in the medium term and protect the planet from devastating climate change in the long term. All three can be done together and all three are urgent. To try to set them against each other as a three-horse race is as confused analytically as it is dangerous economically and environmentally. The urgency of the current crisis is clear, but so also must be the urgency of the climate crisis. The relentless ratchet effect of the build-up of flows of emissions into stocks or concentrations of the greenhouse gases in the atmosphere, together with the ever rising flows, implies that if we are to avoid constantly increasing probabilities of devastating climate change we must see global emissions start to fall before 2020. That means planning and starting action now. Delayed action, even by five or ten years, will see either or both of dangerous increases in risks and sharply greater costs as we try to act in a rushed and ill-considered way later.

We must also, as a world, think, plan and act now to deal with the challenges of adaptation. The climate is changing because of what we have already emitted and will continue to emit in the coming few years. The challenge of adaptation is there for all countries. While more property may be at risk in the rich world, far more lives and livelihoods are at risk in the developing world. We urgently require more detailed analysis of likely impacts in the next few decades and commitment by the rich world to help with the greatly increased resources necessary for economic development and fighting poverty in a more hostile climate.

It should now be clear that we understand enough of the science to see the magnitude of the risks and the scale of the action necessary to reduce them. We can recognise the technologies that can take us both to a more energy-efficient economy and to energy sources which are low-carbon, and we are constantly developing more. We can see how we can stop deforestation. Thus we understand the practicalities of what to do and how to do it; and, given how that understanding has deepened so quickly over the last few years, we can be confident we will learn much more, and reduce costs, along the way.

We also understand the economic policies that can deliver the necessary action in an effective, efficient and equitable way. The costs of the necessary actions are acceptable and manageable in relation to the reduction in risks they bring. Each country will choose its own way forward: there are many ways to construct policy to foster low-carbon growth and different countries will choose different combinations of actions and policies depending on their physical, political, cultural and historical circumstances.

It is crystal clear, however, that this is a global challenge and can be confronted effectively only by concerted action across the world. This will need international collaboration on an unprecedented scale: that is the only way it can work. There are different forms of mutual understandings and institutions that can support such action, but a spirit of internationalism, mutual dependence and shared destiny is fundamental. If we cannot create this collaboration, we will have failed future generations and ourselves.

The meeting of the United Nations Framework Convention on Climate Change at Copenhagen in December 2009, the fifteenth conference of the parties (COP15), is the most important international gathering since the Second World War. The world set itself the task in Bali, at COP13, in December 2007, of reaching an agreement on a successor to the Kyoto Protocol by the end of 2009. If we fail to construct a strong global deal in Copenhagen, we risk years of dangerous delay. This is not like a WTO negotiation in which delay is a loss but we can pick up again roughly where we started – delay means higher concentrations and growing emissions. The starting point for both stocks and flows gets worse. And, further, the confidence in future policy from those in the private sector who will be the investors, and those who will take the practical actions, will be severely damaged. The emerging carbon markets, crucial to necessary incentives, will be undermined. We cannot postpone the construction, agreement and action on a global deal.

As I write, in the autumn of 2008, momentum is growing, particularly with the election of a new president of the United States and the deepening commitment of China and many other countries of the developing world. But we have a long way to go and agreement on a deal of the necessary ambition is far from certain. There will be many voicing confused short-term arguments, and many following their narrow perceptions of short-term self-interest. Sustained political pressure from the people of all countries who understand the nature of the challenge and how to respond is crucial.

Above all, we will need political leadership which is not only thoughtful and measured but also courageous and inspirational. That leadership must set out the compelling scientific and economic case for strong action. It must show not only the severe dangers of a planet in peril, but also that if we act sensibly and strongly, starting now, we can dramatically reduce those risks at reasonable cost. That leadership must be courageous too in confronting the narrow interests which will make a lot of noise and argue for postponement of action, or in some cases for little or no action. It is a time for clarity and strength in both vision and action.

This is indeed an inspirational story. But it is also a practical story – indeed, the *only* practical story. We have a short window of opportunity to turn it into a reality. While it is time for leadership, we must all contribute to the creation of this reality, whether we are from my own world of the university and policy analysis, investors in the new opportunities or people who will change the way we consume. We know what we have to do; the prize is enormous. The citizens and politicians of the world, community by community, nation by nation, will now determine whether we can create and sustain the international vision, commitment and collaboration which will allow us to take this special opportunity and rise to the challenge of a planet in peril.

Notes

Chapter 2

1. Fourier, Joseph (1827), pp. 569–604.
2. Arrhenius, Svante (1896), pp. 237–76.
3. Modern research now deals with probablities of different temperature increases over a large range rather than simple point estimates, since we cannot be sure about many of the key elements in such a complex system as the global climate. This uncertainty is crucial.
4. Total greenhouse gas emissions are measured in CO_2e ('CO_2 equivalent'). It is obtained by aggregating non-CO_2 greenhouse gas flows with those of CO_2 using weights which reflect the respective contribution of each gas to the change in net radiation at the upper troposphere. One gigatonne (Gt) is a billion tonnes. The figures also include all sources of emissions, both energy and non-energy. Deforestation is an important source of non-energy emissions, indeed around 15–20% of the world total. Here these emissions are attributed to the country in which the deforestation occurs. In comparing figures on emissions confusion sometimes arises over CO_2e versus CO_2 only and all sources of emissions versus energy only (see also notes 5, 6, 10, 15).
5. We use Annex 1 in the Kyoto Protocol as defining high-income and non-Annex 1 as developing and the standard UN / World Bank definition of least developed. Country level figures should be treated with some caution. The best source available is from the World Resources Institute (see also note 10). The data for 2000 appear more robust than for other years. Long-time series of CO_2e by country are not available – we use CO_2 here but the essential points on comparisons and trends are clear. For Annex 1 countries as a whole moving from CO_2 to CO_2e adds around a quarter and for non-Annex 1, around a third.
6. The population numbers are for 2004. The 'developed' group referred to in the figure is the Annex 1 group as defined by the UNFCCC, and includes most high- and middle-income countries: Australia, Austria, Belarus, Belgium, Bulgaria, Canada, Croatia, Czech Republic, Denmark, Estonia, Finland, France, Germany, Greece, Hungary, Iceland, Ireland, Italy, Japan, Latvia, Liechtenstein, Lithuania, Luxembourg, Monaco, Netherlands, New

Zealand, Norway, Poland, Portugal, Romania, Russian Federation, Slovakia, Slovenia, Spain, Sweden, Switzerland, Turkey, Ukraine, United Kingdom, United States of America. Some of these, for example, Romania, Turkey and Ukraine, are not amongst the standard World Bank 'high income' group and thus the total at 1.2 billion is higher than the 1 billion in the 'high income' category of the World Bank. 'Developing' in the figure is the 'non-Annex' 1 nations, mostly developing countries, except Brunei and Iraq which are not included. Taiwan is treated as a province of China. LDCs include the poorest countries, in terms of income, resources and vulnerability.

7. See World Development Indicators 2008, World Bank, for a recent reassessment using a revised benchmark of $1.40 per day, taking account of changes in purchasing power parity.

8 'Today' refers to levels around 2005. The annual addition to concentrations over the last four years has been about 2.5 ppm.

9 In 1850 the radiative forcing of non-CO_2 greenhouse gases was close to zero, while of today's 430 ppm CO_2e stock, around 385 is CO_2 (Etheridge, D. M., et al. 1996). Global means and growth rates for current levels constructed using about 70 Climate Monitoring & Diagnostics Laboratory Carbon Cycle-Greenhouse Gases (CMDL CCGG) Sampling Network station data (Conway et al. 2008).

10. Historical emissions data accessible from Climate Action Indicator Tools (CAIT) Database as part of the Climate, Energy and Pollution Program of the World Resources Institute (WRI) (2008).

11. *Energy-related CO_2 emissions by region* (source: World Energy Outlook 2006, IEA, 2006):

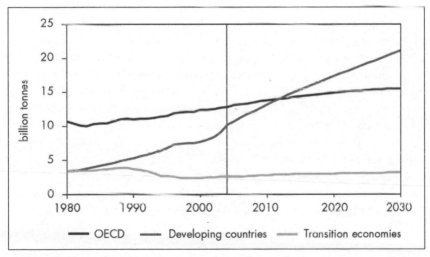

'Transition economies' are the economies of Eastern Europe and the former Soviet Union in transition from a command to a market economy.

12. Nakicenovic, N. and Swart, R. (eds) (2000).

13. The USA emitted approximately 290 Gt CO_2 during the last century (CAIT database, WRI, 2008). In 2004, China emitted 5.4 Gt CO_2. Assuming an average growth of 5.1% year on year (after the Garnaut Climate Change Review, p. 65, box 3.1, Final Report, October 2008), annual emissions for China would grow to approximately 19 Gt CO_2 by 2030, and cumulative emissions for the period 2005–30 would total approximately 289 Gt CO_2.

14. United Nations Food and Agricultural Organisation (FAO) (2005).

15. Kyoto greenhouse gases include CO_2, CH_4, N_2O, PFCs, HFCs, SF_6.

16. One can build more or less elaborate models of these processes, but it is hard to avoid the conclusion that annual increases of concentrations in this century will, in the absence of strong and effective policy, be around these rates.

17. For the mathematically inclined reader, note that the relationship between concentrations and temperature appears to be logarithmic approximately, so that the starting point for the doubling would not change the estimated temperature increase. Also, calculations of long-run equilibrium usually assume the eventual elimination of aerosols.

18. Stainforth et al. (2005), pp. 403–6.

19. The probabilities associated with eventual temperature increases arising from given concentrations assume that aerosols are eventually very low. This would indeed be a likely consequence of cleaner technologies. Aerosols have the effect of diminishing the radiative forcing. At current levels this effect might be equivalent to a reduction of concentrations of around 60 ppm CO_2e.

20. Arntzen, J. W. and Hulme, M. (1996).

21. Coe, M. T. and Foley, J. A. (2001).

22. International Centre for Integrated Mountain Development (2002).

23. Thompson, L. G. et al. (2000).

24. Dyurgerov, M. B. and Meier, M. F. (1997), pp. 379–91.

25. Singh, P. et al. (1994).

26. Kocin, P. et al. (1998), pp. 47–54.

27. NCDC (1998).

28. The numbers of those who are deniers on the basis of analytical science are very small in relation to science as a whole. There are, of course, many people in the coffee houses, bars, and barber shops of the world who are casual deniers of the science.

Chapter 3

1.   'The Globalist', 9 June 2008.
2.   See Stern (2007), Figure 8.2.
3.   See, for example, ibid., Figure 8.1.
4   Ibid., Figure 8.4.

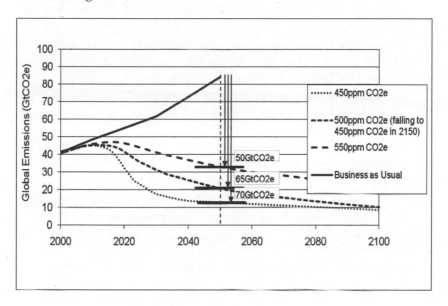

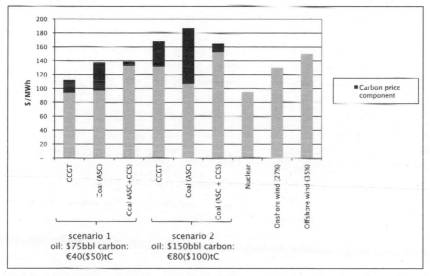

Levelised cost is current plus capital cost where the latter is converted to a flow.

5.   Care is necessary in comparing figures on breakdowns across sectors. In addition to variations across countries, sometimes emissions figures are given as energy only and sometimes CO_2 only. We generally refer to all sources and all greenhouse gases unless otherwise stated.

6.   See www.cait.wri.org/downloads/cait_ghgs.pdf.

7.   With nuclear, it is always necessary to make appropriate assumptions about decommissioning costs. Our figures are drawn from the UK government 'Energy Challenge Energy Review Report 2006', Department of Trade and Industry (July 2006), which takes into account the full life-cycle carbon emissions of nuclear, including decommissioning, against a full range of generating technologies.

8.   We use an exchange rate of €1 = approximately $1.30.

9.   *Costs of Electricity for Different Technologies and Price*:

10.   See Bundestministerium für Umwelt, Naturschutz und Reatkorsicherheit (31 July 2008).

11.   Nanobatteries employ a scale of minuscule particles that measure less than 100 nanometres, roughly a hundred times smaller than particles used in traditional Li-Ion technology. Nanobatteries have a higher energy density and recharge faster and more often than chemical batteries.

12.   See www.caac.gov.cn.

13.   See also Aghion and Howitt (1990) and (2009) and Aghion (2007) for a recent theoretical account in the context of climate change; and Schumpeter (1942).

14.   Fankhauser et al. (2008) review the literature on the opportunities for energy and jobs arising from strong policy on climate change. They examine short-term employment effects, medium-term effects on the economy as a whole and longer-term induced growth effects. While these types of analyses are at an early stage, there are indications that, with good policy and a prompt beginning so that adjustments can take place in a measured way, all three efforts are likely to be positive.

15.   In indicating the crude percentages as cost estimates, I have been fairly cautious with assumptions on technical progress and 'learning from experience'. Once new technologies are adopted, costs as a percentage of GDP may fall, and on a number of occasions I have rounded approximations up, rather than down. On the other hand, we have to allow for a world which does not always choose good policies or lowest cost options. I would suggest, therefore, that these calculations are neither over- nor under-optimistic from the point of view of costs in relation to GDP. And looking back, I think that, on this basis, the Stern Review estimates of costs of action were in the right ballpark.

16. Models, be they bottom-up or top-down, vary greatly in their assumptions and the variations lead to different results. Key assumptions include: levels and growth rates of emissions; flexibility between sectors, technologies, gases, countries; and the rate of discovery of new technologies.

17. This would mean that a point on the curve would actually represent the extra cost of increasing abatement by one tonne from the abatement level indicated on the horizontal axis – thus would involve the assumption that the cheaper methods are indeed chosen ahead of the more expensive and the technological and cost descriptions are correct.

18. IEA World Energy Outlook 2008 suggests a price of over $100 per barrel in real terms for the next two decades (looking beyond the current world slowdown).

19. British Wind Energy Association press release (26 January 2005).

20. British Wind Energy Association briefing sheet (October 2005).

21. Global Wind Energy Council Report (2007).

Chapter 4

1. www.ipcc.ch/pdf/technical-papers/climate-change-water-en.pdf.

2. Before leading the Stern Review, I led the writing of the Report of the Commission for Africa. See this, and background papers, for a discussion of African agriculture: www.commissionforafrica.org.

3. See the Report of the Commission for Africa. On African population, see www.un.org/esa/population/unpop.htm.

4. See, for example, Klaus (2008) and Lawson (2008).

5. See, for example, the assessment in the 'Summary for Policymakers' in the Contribution of Working Group II to the AR4, www.ipcc.ch/pdf/assessment-report/ar4/wg2/ar4-wg2-spm.pdf.

6. We might measure improved quality by the reduction in cost necessary to achieve a given outcome.

7. climateprediction.net, the world's largest climate forecast experiment, is based on distributed computing power. Stainforth et al. (2005), provide an excellent discussion of its first scientific results, available at www.climateprediction.net/science/pubs/nature_first_results.pdf.

8. United Nations Development Programme (2007), p. 13.

9. Bettencourt, S. et al., (2006)

10. Brown, Jeff (2005).

11. UN database on local coping strategies. See www.maindb.unfccc.int/public/adaptation/adaptation_casestudy.pl?id_project=55).

12. ONERC (2005).

13. UN database on local coping strategies, www.maindb.unfccc.int/public/adaptation/adaptation_casestudy.pl?id_project=149).

14. The CGIAR is a worldwide group of research institutes, each of which might have a focus on particular crops, such as the International Rice Research Institute (IRRI) in Manila, or a particular area, such as the International Crop Research Institute for the Semi-Arid Tropics (ICRISAT) in Hyderabad, India, or on a particular issue, such as the International Food Policy Research Institute (IFPRI) in Washington DC.

15. See www.cset.iastate.edu/research-projects/bio-char.html.

16. See www.nytimes.com/2007/10/16/nyregion/16insurance.html?ei=5089&en=7c5a5d3525adfb75&ex=1350187200&partner=rssyahoo&emc=rss&pagewanted=all.

17. See www.abi.org.uk/Newsreleases/viewNewsRelease.asp?nrid=16415.

18. See www.rediff.com/money/2008/feb/28ins.htm. For some further references see chapter 20 of the Stern Review.

19. United Nations Development Programme, op. cit., p. 166.

20. Very early estimates on possible costs to infrastructure were provided by the World Bank in 2005. The next World Development Report of the World Bank, to be published in 2009, will provide further analysis. There is some discussion of the issues in Chapter 20 of the Stern Review. The *Human Development Report 2007/2008* of the United Nations Developmant Programme published in November 2007 provides some analysis, as does the report of Working Group 3 of the IPCC published in 2007. The informational basis of and strategic plans for adaptation and its costings should be a priority for further work. See Metz et al. (2007), and: www.ipcc.ch/ipccreports/ar4-wg3.htm.

21. The UNFCCC Secretariat report, Investment and Financial Flows to Address Climate Change estimates extra costs of development for 2030 are $28–67 billion although they exclude strengthening poverty reduction strategies. For this reason and in the light of the arguments and examples in the text this may be on the low side. Available at unfcc.int/files/cooperation_and_support/financial_mechanism/financial_mechanism_gef/application/pdf/dialogue_working_paper_8.pdf.

22. Bates et al. (2008). See www.ipcc.ch/pdf/technical-papers/climate-change-water-en.pdf.

23. This target was originally agreed in UN General Assembly Resolution 2626 in 1970. It was reaffirmed with reference to the Millennium Development Goals in the Monterrey UN Financing for Development meeting of 2002, and in June 2005 the EU set this target for achievement by 2015.

24. United Nations Development Programme, op. cit., p. 25.

25. The full declaration is at www.oecd.org/dataoecd/11/41/34428351.pdf.

26. United Nations Development Programme, op. cit., (2007), p. 166.

Chapter 5

1. See, for example, Nordhaus (2008). This statement was surprising as he
    is a scholar and a gentleman. He is simply misguided and misleading on
    the key economic issues discussed in this chapter, as we shall show.

2. This has been proposed, for example, by Weitzman (2007). Nordhaus
    (2007) proposes a similar rate of 1.5%.

3. Notice that while we might treat equally an increase in consumption to
    all those alive today, who have equal consumption today, it would still
    be the case that, taking two people with the same lifetime pattern, the
    one born later would have lower weight – any increase to them at a
    particular age would have lower value.

4. There is another type of problem which can arise with pure-time
    discounting concerning 'time inconsistency'. Suppose that instead of
    discriminating by birth as in the above example, we treat all people alive
    at a given time in the same way, irrespective of age. Then with pure-
    time discounting the ethical status of an individual changes at birth (we
    counted him/her at less value than current generations prior to birth but
    equal after birth). Then we might regard past decisions as in error in
    attaching too little weight to that individual.

5. Stern, N., (1977), pp. 209–54; and Stern, N., (2008) pp. 1–37.

6. The idea is similar to the standard one in economics that if people
    choose between apples and pears facing fixed prices, then they will go on
    adjusting their purchases up to the point where their marginal valuation
    of an extra apple relative to an extra pear is equal to the price ratio they
    face in the market. Here, the argument is that interest rates indicate the
    marginal valuation of an extra unit of consumption in the future relative
    to a unit today. For example, if the real interest rate between today and
    tomorrow is 5%, then a unit today becomes 1.05 units tomorrow. This is
    taken as an indication that 'society' considers a unit today to be 'worth'
    1.05 units tomorrow.

7. Unless, of course, the damages are indefinitely large. Weitzman (2007b)
    has emphasised this important point.

8. More generally, there are many types of income in an economy with a
    variety of public, private and public-private activities, with taxation,
    with risk, and so on. These include public and private consumption or
    income, public and private investment, and so on. In an economy with
    limited tax and transfer instruments, different forms of income will have
    different social values and different associated discount rates.

9. This argument has been very well made by Sterner, T., and Persson, U.
    M., (2007), and Guesnerie, R., (2004), pp. 363–82.

10. See, for example, Lomborg. B., (2007).

11.   See, for example, Lomborg's very confused analyses. The mistake comes from a failure to understand that, in imperfect economies, different forms of income can have different social values.

12.   The reader whose appetite for the economics of discounting is not already sated may want to look at my 2008 Richard T. Ely Lecture (Stern [2008]) for further analysis of these issues.

13.   The costs of inaction might equivalently be referred to as the 'benefits of action' – the costs from climate change that would be incurred under business as usual, but which would be avoided by the set of actions to combat climate change under consideration. However, I prefer the term 'costs of inaction', as it more clearly conveys the idea that doing nothing is itself a choice, and, moreover, one that will involve very large costs for society. It also focuses on choices between strategies rather than the 'benefit-cost' language which is readily associated with investment projects which are small relative to the world as a whole.

14.   Formally, the expression takes a form like *Damages* = *A (Temperature Increase)* γ, where *A* and γ are chosen by the modeller.

15.   Within the broad range of 5–20%, there is a 'central case' with damages around 10–11% of GDP per annum. The lower number of 5% is associated with a very narrow 'market-goods only' view of damages. The higher number of 20% includes allowance for a greater climate sensitivity than is assumed in the modelling (which probably was indeed, as discussed earlier, on the low side, at least in terms of probabilities of higher temperatures) and, additionally, brings in concerns about intra-generational distribution. These considerations point to a focus on the upper part of the 5–20% range. The ethical assumptions used in the base case had, first, low rates of pure-time discounting (around 0.1% per annum). This was for the reasons described above – that is, we did not believe that there was a convincing case for pure-time discounting, except in order to account for the possibility that the human race might cease to exist (for some reason other than human-induced climate change). Second, the base case had distributional values which focused only on income (or consumption) and portrayed the social value of an extra unit of income to a person as being worth half that of an extra unit to a person with half the wealth. This second assumption is an example of the motivation for discounting which puts lower weights on future increments because it is assumed that future generations will have higher incomes than ourselves. The Stern Review provided a range of examples on distribution values and pure-time discount rates and the calculations are further discussed in Stern (2008).

16. They include, for example, Mendelsohn, R. et al. (2000), pp. 553–69; Nordhaus, William D. and Boyer, Joseph G. (2000); Tol, Richard S. J. (2002), pp. 135–60.

17. See, for example, Mendelsohn et al., ibid. and Tol, ibid.

18. See, for example, Nordhaus (2007 and 2008).

19. Weitzman (2007 and 2007b) has shown that there are serious analytical difficulties involved in modelling the potentially very large risks that are inherent in climate change.

20. A detailed, somewhat formal and mathematical discussion of the theory of policy in imperfect and multi-good models may be found in my chapter (with Jean Drèze) of the *Handbook of Public Economics* edited by Auerbach and Feldstein. Drèze (1987), pp. 909–89.

21. Robert Hefner III's book *The Grand Energy Transition*, for example, describes what this transition might involve, and compares it to previous transitions in humanity's use of energy sources.

Chapter 6

1. See, for example, Ariely (2008) on how the very short run can get 'excessive weight in decision-making'.

2. For the record, Dion took this slogan to Canada's 2008 elections and was defeated.

3. See Anderson, J. F. and Sherwood, T. (2002) or Hodges, H. (1997).
    A further point to note is that businesses are often positioned well within the production possibility frontier. As De Canio points out, 'realising even a fraction of these potential gains would outweigh the most pessimistic estimates of the cost of climate stabilisation strategy.' De Canio, S. J. (July 1999), pp. 279–95.

4. See, for example, 'Including aviation in the EU Emissions Trading Scheme – an estimate of the potential windfall profit' (December 2006), WWF, www.wwf.org.uk/filelibrary/pdf/windfalls.pdf.

5. See 'Regional Greenhouse Gas Initiative: an initiative of the Northeast and Mid-Atlantic States of the U.S.' (2008), RGGI Inc., www.rggi.org.

6. 'IEA Energy R&D Statistics' (2008).

7. Stern (2007), p. 401.

8. 'IEA Energy R&D Statistics', op. cit.

9. See Stern (2007), Chapter 16.

10. Ibid., p. 403.

11. This estimate reflects the fact that total GDP in high-income countries is around $35 billion (in current prices). See, for example, the World Development Indicators of the World Bank (2008), www.world

bank.org/WBSITE/EXTERNAL/DATASTATISTICS/O,,contentMD
K:21725423~pagePK:64133150~piPK:64133175~theSitePK:239419,00.html).

12. See IEA *World Energy Outlook 2008*, Chapter 2, International Energy
Agency, Paris; and Stern (2007), pp. 421–2.

13. See, for example, Sunstein and Thaler (2008).

14. See 'The Eddington Transport Study: The case for action: Sir Rod
Eddington's advice to Government' (December 2006), Executive
Summary, p. 5, UK Department for Transport, www.dft.gov.uk/
162259/187604/206711/executivesummary.

15. Ibid., p. 6.

16. Ottmar Edenhofer's research-in-progress suggests that if CCS is unavail-
able and there is not strong technological progress in other renewable
technologies, costs of action might double (Edenhofer et al. [2008],
'Robust Options for Decarbonisation', in forthcoming 2009 book with
the working title *Global Sustainability – A Nobel Cause*; and 'Special Issue of
the Energy Journal – The Economics of Low Stabilisation', also
forthcoming).

Chapter 7

1. Farrell, Diana and Halstead, Ted (21 June 2007).

2. Bressand, Florian et al. (May 2007).

3. 'Home Insulation and Glazing' (2008), *Energy Saving Trust*,
http://www.energysavingtrust.org.uk/Home-improvements/Home-
insulation-glazing.

4. 'Inventory of US greenhouse gas emissions and sinks: 1990–2006' (April
2008), *Environmental Protection Agency*, www.epa.gov/climatechange/
emissions/usinventoryreport.html.

5. See Goodall (2007).

6. See 'Another Green Revolution', *Technology Quarterly* in *The Economist*, 4
September 2008.

7. It excludes deforestation emissions. For the study, see Eshel, Gidon and
Martin, Pamela (April 2006), pp. 1–17.

8. See also 'Another Green Revolution' (note 6). At present the cost of
Light Emitting Diode (LED) lamps is high but it is falling rapidly. The
example shows the advantages of both technical progress and a source
of electricity for poor households.

9. As a part-time adviser to HSBC's group chairman, Stephen Green, I am
particularly aware of some of the information and analysis they
assemble and have drawn on it for some of the examples of this chapter.

10. This report was based on research conducted with 9,000 consumers in

Brazil, China, France, Germany, Hong Kong, India, Mexico, UK and USA in April 2007. 'HSBC Climate Confidence Index 2007', HSBC Climate Partnership, www.hsbc.com/1/PA_1_1_S5/content/assets/newsroom/hsbc_ccindex_p8.pdf.

11. The Climate Group is an international organisation working with businesses across the globe to raise awareness and offer practical recommendations for individuals, businesses and communities on the path towards a low-carbon economy.

12. A recent report found that 'consumer emissions', which includes emissions from goods and services consumed by UK residents, wherever they come from, are 37% higher than the national total reported to the UNFCCC ('Development of an Embedded Carbon Emissions Indicator: A Research Report to the Department of Environment, Food and Rural Affairs', July 2008, Stockholm Environment Institute and the University of Sydney, www.randd.defra.gov.uk).

13. 'Solar Power' (2008), *Project Highlights*, The Energy and Resources Institute, North America, www.terina.org/index.php?option=com_content&task=view&id=20.

14. For this and other examples, see www.fandc.com/insurance.

15. See: www.worldwildlife.org/climate/item3799.html.

16. From Dr Jonathan Pershing's Testimony to the US House of Representatives Subcommittee on Energy and Air Quality.

17. The survey was conducted for the BBC World Service by the international polling firm GlobeScan together with the Program on International Policy Attitudes (PIPA) at the University of Maryland.

18. 'HSBC Climate Confidence Index 2007' (2007), HSBC Climate Partnership, www.hsbc.com/1/PA_1_1_S5/content/assets/newsroom/hsbc_ccindex_p8.pdf.

19. 'A Look at Emissions Targets' (2008), *Pew Center on Global Climate Change*, www.pewclimate.org/what_s_being_done/targets.

20. 'How Companies Think About Climate Change: A McKinsey Global Survey', *McKinsey Quarterly*, February 2008.

21. Newing, R., (9 October 2008).

22. See www.environmentalleader.com/2008.

23. See Esty and Winston (2006).

24. Sigma/Swiss Re, 'Annual Natural Catastrophes and Man-made Disasters reports, Oliver Wyman analysis' (2005), *Financial Risks of Climate Change*, Association of British Insurers.

25. 'Cleaning Up – Focus on Global Venture Capital and Private Equity Investment in Clean Energy', New Energy Finance, 10 July 2006, www.newenergyfinance.com/NEF/HTML/Press/2008-01-25-_PR_Investment_in_clean_energy.pdf.

26. Makower, J., et al. (March 2008).

27. For these and other figures, see New Carbon Finance at: http://www.newcarbonfinance.com.

28. Figures within the report showed that China was already the leading producer in terms of installed renewable generation capacity. It has the world's largest hydroelectricity capacity since the controversial Three Gorges project began producing electricity, and the fifth largest fleet of wind turbines on the planet. Although its installed capacity of photovoltaic (PV) panels is still relatively low, it is already a leading manufacturer of them.

29. See www.cdproject.net.

30. Fleming, Peyton (6 March 2008).

31. Worsnip, Patrick (10 October 2008).

32. The group was the Corporate Leaders Group on Climate Change. See www.cpi.cam.ac.uk/bep/clgcc.

33. 'Building a Low Carbon Indian Economy', Confederation of Indian Industry Discussion Paper, (January 2008), www.cii.in/documents/building_lowcarbon08.pdf.

34. 'Business Leaders in Davos Call for a Clear Post-2012 Climate Change Agreement', Press Release, 25 January 2008, Joint International Emissions Trading Association, UNFCCC, World Business Council for Sustainable Development, WEF, Vattenfall AB, International Energy Agency (IEA), www.ieta.org/ieta/www/pages/getfile.php?docID=2867.

35. 'About the Project' (2008), Prince's Rainforest Project, www.princes rainforestsproject.org/about-us/about-the-project.

36. See www.c40cities.org.

37. Ibid.

38. The petitioners were the states of California, Connecticut, Illinois, Maine, Massachusetts, New Jersey, New Mexico, New York, Oregon, Rhode Island, Vermont and Washington, the cities of New York, Baltimore, and Washington DC, the territory of American Samoa, and the organisations Center for Biological Diversity, Center for Food Safety, Conservation Law Foundation, Environmental Advocates, Environmental Defense, Friends of the Earth, Greenpeace, International Center for Technology Assessment, National Environmental Trust, Natural Resources Defense Council, Sierra Club, Union of Concerned Scientists and US Public Interest Research Group. See Supreme Court of the United States Syllabus, 'Massachusetts versus the Environmental Protection Agency et al.', (2 April 2007), www.supremecourtus.gov/opinions/06pdf/05-1120.pdf.

Chapter 8

1. This chapter draws heavily on the paper by Stern, Nicholas, 'Key elements of a global deal on climate change', launched on the London School of Economics website on 30 April 2008 (see www.lse.ac.uk and search for 'Stern Global Deal'). An excellent group of people and organisations were involved in putting that paper together, in which detailed acknowledgements are provided.

2. See Stern (2007), Figure 8.1.

3. Over the indefinite future, stabilisation requires getting emissions down to 5 or 6 Gt CO_2e per annum, which could continue at that level indefinitely. See Stern (2007), Chapters 1 and 8.

4. The patterns of flows over time does have relevance too since the amount absorbed in a period depends on the amount emitted – see Stern (2007), Chapter 8.

5. The discussion of CDM here leans heavily on that of Stern, 'Key elements of a global deal on climate change', op. cit., and draws on the work of Samuel Fankhauser.

6. Preliminary results from the GLOCAF Model of the UK's Office of Climate Change. See results in Stern, op. cit. Work is still in progress and results depend significantly on the scenario (e.g. structure of the carbon market).

7. See Ellis, J. and Kamel, S. (May 2007).

8. On 23 January 2008, the European Commission adopted the 'Climate and Energy' package which includes legislative proposals for Member States emission-reduction targets, the geological storage of CO_2, the Community GHG emission allowance trading system and the use of energy from renewable sources.

9. See Stern (2007), Chapter 10.

10. According to 'Sustaining Forests: A Development Strategy', World Bank, Washington DC, 2004. Box 11, page 16.

11. See, for example, Grieg-Gran, M. (2006), Obersteiner, M. (2006) and Blaser, J. and Robledo, C. (2007).

12. Eliasch, J. (October 2008).

13. See the recent important study by Sukhdev, Pavan, 'The Economics of Ecosystem and Biodiversity' (May 2008), presented at the 9th Conference of the Parties of the Convention on Biological Diversity, Bonn.

14. See Gallagher, E., (July 2008).

15. See Torras, M. (2006), pp. 283–97.

16. See www.iwokrama.org/home.htm, which details the Iwokrama Rain Forest Program implemented in Guyana by the UNDP beginning in 1991.

17. The International Energy Agency has suggested twenty plants. Given the variation of geology across five continents, the different types of technology (e.g. capture pre- and post-combustion), different legal structures and the fact that a number of countries would want to see examples on their own territory, twenty seems low to me.

18. Stern (2008).

19. See the discussion in Chapter 4.

20. See for example, 'World Development Indicators', World Bank, Washington DC, (2008).

Chapter 9

1. For a good summary of Zhu Rongji's position on harmony between economic development and resource and environmental protection, see the speech to the World Summit on Sustainable Development in September 2002. www.china-un.ch/eng/qtzz/wtojjwt/t85656.htm.

2. See Houser, T., et al. (May 2008).

3. Tamirisa, N., et al. (April 2008).

4. On 23 January 2008, the European Commission put forward a far-reaching package of proposals that will deliver on the EU's commitments to fight climate change and promote renewable energy up to 2020 and beyond. The text is available at www.ec.europa.eu/commission_barroso/president/focus/energy-package-2008/index_en.htm#key.

5. For example, the new standard 'Euro 5', which will become effective on 1 September 2009, mandates an 80% cut in the emission limit for particulate matters from diesel cars sold in the EU, and makes the introduction of particle filters for diesel cars obligatory. The standard 'Euro 6' will set 68% lower emission limits for NOx emissions from diesel cars and will come into force in 2014.

6. From David Grey, a senior water analyst at the World Bank. See also the Chinese Academy of Science, suggesting that in western China glaciers shrank by up to 20% in the last forty years, www.news.mongabay.com/2007/0713–china.html.

7. His full speech can be found at www.pmindia.nic.in/lspeech.asp?id=700.

8. Antweiler, W. et al. (2001), pp. 877–908; and Copeland, B.R. and Taylor, M. S. (2004), pp. 7–71.

9. Eliasch, Johan and Peace, Janet and Weyant (April 2008), (October 2008).

10. Farrell, Alexander E. et al. (29 May 2007).

11 The list of functions includes those specified in Section 8 of Stern, Nicholas, 'Key elements of a global deal on climate change' (2008b), the London School of Economics found at www.lse.ac.uk. I am grateful to Nick Butler for discussion and suggestions on these issues.

Bibliography

Aghion, P., 'Environment and Endogenous Technical Change' (Mimeo, Harvard, December 2007)

Aghion, P., and Howitt, P., *Endogenous Growth Theory* (MIT Press, Cambridge, Mass., 1990)

Aghion, P., and Howitt, P., *The Economics of Growth* (MIT Press, Cambridge, Mass., 2009)

Anderson, J. F. and Sherwood, T., 'Comparison of EPA and Other Estimates of Mobile Source Rule Costs to Actual Price Changes', SAE Technical Paper 2002–01–1980, (Society of Automotive Engineers, Warrendale PA, 2002)

Antweiler, W., Copeland, B. R. and Taylor, M. S., 'Is free trade good for the environment?' *American Economic Review,* (2001)

Arntzen, J. W. and Hulme, M., 'Climate Change and Southern Africa: An Exploration of Some Potential Impacts and Implications for the Region', (Climatic Research Unit, University of East Anglia, 1996)

Ariely, D., *Predictably Irrational: The Hidden Forces That Shape Our Decisions* (HarperCollins, London, 2008)

Arrhenius, S., 'On the Influence of Carbonic Acid in the Air Upon the Temperature of the Ground', *Philosophical Magazine 41,* (1896)

Bates, B. C., Kundzewicz, Z. W., Wu, S. and Palutikof, J. P. (eds), IPCC, *Climate Change and Water: Technical Paper of the Intergovernmental Panel on Climate Change* (Geneva, 2008)

Beckerman, W. and Hepburn, C., 'Ethics of the Discount Rate in the Stern Review on the Economics of Climate Change', *World Economics,* 8(1), (2007)

Bettencourt, S., Richard, C., Freeman, P. et al., 'Not if but when: Adapting to natural hazards in the Pacific Islands region, A policy note,'(siteresources.worldbank.org./INTPACIFICISLANDS/Resources/Natural-Hazardsreport.Pdf), (World Bank, Washington DC, 2006)

Blaser, J. and Robledo, C., 'Initial Analysis of the Mitigation Potential in the Forestry Sector', UNFCCC Secretariat, (www.unfccc.int/cooperation_and_support/financial_mechanism/financial_mecha nism_gef/items/4054.php), (2007)

Bressand, F., et al., 'Curbing global energy demand growth: The energy productivity opportunity', McKinsey *Global Institute*, (http://www.mckinsey.com/mgi/publications/Curbing_Global_ Energy/index.asp), (May 2007)

British Wind Energy Association 'Government action on renewables undermines fine intentions on climate change', (www.bwea.com/media/news/rates.html), (26 January 2005)

British Wind Energy Association, 'Wind energy and the UK's 10% Target', www.bwea.com/pdf/briefings/10target2005.pdf), (October 2005)

Brown, J., 'Cold region design: high-altitude railway designed to survive climate change', *Civil Engineering*, (2005)

Bundestministerium für Umwelt, Naturschutz und Reatkorsicherheit, *Erneuerbare Energie in Zahlen* (www.erneuerbare-energien.de/files/erneuerbare_energien/downloads/application/pdf/broschuere_ee _zahlen.pdf), (2008)

Climate Action Indicator Tools (CAIT) Database, World Resources Institute (WRI), (www.cait.wri.org), (2008)

Coe, M. T. and Foley, J. A., 'Human and natural impacts on the water resources of the Lake Chad basin', *Journal of Geophysical Research*, 106, (2001)

Conway, T. J., Lang, P. M. and Masarie, K. A., Atmospheric Carbon Dioxide Dry Air Mole Fractions from The NOAA ESRL Carbon Cycle cooperative Global Air Sampling Network, (www.cmdl.noaa.gov /ccg/co$_2$/flask/event/, (1968-2007), version: (2008-07-24.)

Copeland, B.R. and Taylor, M. S., 'Trade, growth and the environment', *Journal of Economic Literature*, (2004)

De Canio, S. J., 'Estimating non-environmental consequences of greenhouse gas reductions is harder than you think', *Contemporary Economic Policy*, Vol. 17, No. 3, (July 1999)

Drèze, J. and Stern, N., 'The theory of costbenefit analysis' in Auerbach, A. J. and Feldstein, M. (eds), *Handbook of Public Economics*, edn 1, Vol. 2, (Elsevier, Amsterdam, 1987)

Dyurgerov, M. B. and Meier, M. F., 'Mass balance of mountain and subpolar glaciers: A new global assessment for 1961–1990', *Arctic and Alpine Research*, (1997)

Eliasch, J., 'Climate Change: Financing Global Forests', *The Eliasch Review*, (UK Office of Climate Change, October 2008)

Ellis, J. and Kamel, S., 'Overcoming barriers to Clean Development Mechanism projects', (OECD and UNEP/RISOE, May 2007)

Enkvist, P., Nauclér, T. and Rosander, J., 'A cost curve for greenhouse gas reduction', *McKinsey Quarterly* 1, (2007)

Eshel, G. and Martin, P., 'Diet, Energy and GlobalWarming', *Earth Interactions* (10:9), (April 2006)

Esty, D. and Winston, A., *Green to Gold: How Smart Companies Use Environmental Strategy to Innovate, Create Value and Build Competitive Advantage* (Yale University Press, New Haven, 2006)

Etheridge, D. M., Steele, L. P., Langenfelds, R. L., Francey, R. J., Barnola, J. M. and

Morgan, V. I., 'Natural and anthropogenic changes in atmospheric CO_2 over the last 1,000 years from air in Antarctic ice and firn.', *Journal of Geophysics Research*, (1996)

Farrell, A. E. and Sperling, D., 'A Low-Carbon Fuel Standard for California Part 1: Technical Analysis', (UC Berkeley Transportation Sustainability Research Center, 29 May 2007)

Farrell, D. and Halstead, T., 'US Must Warm to Energy Efficiency', *Financial Times*, (21 June 2007)

Fankhauser, S., Sehlleier, F. and Stern, N., 'Climate change, innovation and jobs', *Climate Policy 8*, (2008)

Fleming, P., 'Investors File Record Number of Global Warming Resolutions with U.S. Companies', Ceres News, (www.ceres.org/NETCOMMUNITY/Page.aspx?pid=854&srcid=421), (6 March 2008)

Fourier, J., 'Mémoire sur les Températures du Globe Terrestre et des Espaces Planétaires', *Mémoires de l'Académie Royale des Sciences 7*, (1827)

Friedman, T., *Hot, Flat and Crowded: Why the World Needs a Green Revolution – and How We Can Renew Our Global Future* (Farrar, Straus & Giroux, New York, 2008)

Gallagher, E., 'The Gallagher Review on indirect effects of biofuels', (UK Renewable Fuels Agency, July 2008)

Garnaut, R., 'Will Climate Change Bring an End to the Platinum Age?', paper resented at the inaugural S. T. Lee Lecture on Asia & The Pacific, (Australian National University, Canberra, 29 November 2007)

Garnaut, R., *The Garnaut Climate Change Review*, (Cambridge University Press, Cambridge 2008)

The Globalist, (www.theglobalist.com), (9 June 2008)

Global Wind Energy Council, *Global Wind Energy Council Report*, (www.gwec.net/fileadmin/documents/test2/gwec-08-update_FINAL.pdf), (2007)

Goodall, C., *How to Live a Low-Carbon Life: The Individual's Guide to Stopping Climate Change* (Earthscan Publications, London, 2007)

Grieg-Gran, M., 'The Cost of Avoiding Deforestation', (International Institute for Environment and Development, 2006)

Guesnerie, R., 'Calcul économique et développement durable', *Revue économique,* (2004)

Hodges, H., 'Falling prices, cost of complying with environmental regulations almost always less than advertised', (www.epinet.org/briefingpapers/bp69.pdf), (Economic Policy Institute, 1997)

Houghton, J. T., Ding, Y., Griggs, D. J., Noguer, M., van der Linden, P.J. and Xiaosu, D. (eds), IPCC, *Climate Change 2001: The Scientific Basis: A Contribution of Working Group I to the Third Assessment Report of the Intergovernmental Panel on Climate Change* (CUP, Cambridge, 2001)

Houser, T., Bradley, R., Childs, B., Werksman, J., Heilmay, R., 'Leveling The Carbon Playing Field: International Competition and US Climate Policy Design', (Peterson Institute for International Economics and World Resources Institute, Washington DC, May 2008)

ICIMOD, 'Inventory of Glaciers, Glacial Lakes, and Glacial Lake Outburst Floods, Monitoring and Early Warning Systems in the Hindu Kush-Himalayan Region – Bhutan, International Centre for Integrated Mountain Development (ICIMOD) and United Nations Environment Programme', (www.rrcap.unep.org/issues/glof/), (2002)

IEA, *World Energy Outlook 2006*, Paris, 2006; *World Energy Outlook 2007*, Paris, (2007)

IEA, *IEA Statistics*, 'IEA Energy R&D Statistics', (www.iea.org/textbase/stats/rd.asp), (Paris, 2008)

Klaus,V., *Blue Planet in Green Shackles* (Competitive Enterprise Institute, Washington DC, 2008)

Knight, F., *Risk, Uncertainty and Profit* (Houghton Mifflin, Boston, 1921)

Kocin, P., Gartner, W. and Graf, D., 'The 1996–97 snow season', *Weatherwise* (1998)

Lawson, N., 'Lecture on the Economics and Politics of Climate Change – An Appeal to Reason', Centre for Policy Studies, November 2006, and *An Appeal to Reason: A Cool Look at Global Warming* (Duckworth, London, 2008)

Lomborg, B., *Cool It: The Skeptical Environmentalist's Guide to Global Warming* (Knopf, New York, 2007)

Makower, J., Pernick, R., Wilder, C., 'Clean Energy Trends 2008', *Clean Edge*, (March 2008)

Mendelsohn, R., Morrison,W., Schlesinger, M. and Andronova, N. (2000), 'Country-Specific Market Impacts of Climate Change', *Climatic Change*, 45(3–4)

Metz, B., Davidson, O. R., Bosch, P. R., Dave, R. and Meyer, L. A.(eds), IPCC, *Climate Change 2007: Mitigation: A Contribution of Working Group III to the Fourth Assessment Report of the Intergovernmental Panel on Climate Change* (CUP, Cambridge, 2007)

Nakicenovic, N. and Swart, R. (eds), IPCC, *Emissions Scenarios: A Special Report of Working Group III of the Intergovernmental Panel on Climate Change* (CUP, Cambridge, 2000)

NCDC, 'Topweather and climate stories of 1998. 6 January 1999 Climate Prediction Center', (www.ncdc.noaa.gov/ol/climate/research/1998/ann/top-99.html) (1998)

Newing, R, 'Vehicle sharing - an idea that carries weight in the cause of sustainability', *Financial Times*, (9 October 2008)

Nordhaus, W. D., and Boyer, J. G., *Warming the World: Economic Models of Global Warming*, (MIT Press, Cambridge, Mass., 2000)

Nordhaus, W. D., 'A Review of the Stern Review on the Economics of Climate Change' *Journal of Economic Literature*, (2007)

Nordhaus, W. D., *A Question of Balance* (Yale University Press, New Haven, 2008)

Obersteiner, M., 'Economics of Avoiding Deforestation', (International Institute for Applied Analysis, Austria, 2006)

ONERC, 'Un climat à la derive: comment s'adapter, Rapport de l'ONERC au Premier Ministre et au Parlement, (www.ecologie. gouv.fr/IMG/pdf/Strategie_Nationale_2.17_Mo-2.pdf.), (Observatoire National sur les Effects du Rechauffement Climatique, Paris, 2005)

Peace, J. and Weyant, J., 'Insights Not Numbers: The Appropriate Use of Economic Models', (Pew Center on Global Climate Change, October 2008)

Rawls, J., *Theory of Justice* (OUP, Oxford, 1973)

Schumpeter, J., *Capitalism, Socialism and Democracy* (Harper Perennial, New York, 1942)

Singh, P., Jain, S. K., Kumar, N. and Singh, U. K., 'Snow and Glacier Contribution in the Chenab River at Akhnoor', CS (AR)-131, (National Institute of Hydrology, Roorkee, India, 1994)

Solomon, S., Qin, D., Manning, M., Chen, Z., Marquis, M., Averyt, K. B., Tignor, M. and Miller, H. L. (eds), IPCC, *Climate Change 2007: The Physical Science Basis: A Contribution of Working Group I to the Fourth Assessment Report of the Intergovernmental Panel on Climate Change* (CUP, Cambridge, 2007)

Stainforth, D., Aina, T., Christensen, C., Collins, M., Faull, N., Frame, D. J., Kettleborough, J. A., Knight, S., Martin, A., Murphy, J. M., Piani, C., Sexton, D., Smith, L. A., Spicer, R. A., Thorpe, A. J. and Allen, M. R., 'Uncertainty in predictions of the climate response to rising levels of greenhouse gases', *Nature* 433, (January 2005)

Stern, N., 'The Marginal Valuation of Income' in *Studies in Modern Economic Analysis: The Proceedings of the Association of University Teachers of Economics, Edinburgh 1976*, Artis, M. J. and Nobay, A. R. (eds), (Blackwell, Oxford, 1977)

Stern, N., *The Economics of Climate Change: The Stern Review* (CUP, Cambridge, 2007)

Stern, N., 'The Economics of Climate Change', Richard T. Ely Lecture (*American Economic Review: Papers&Proceedings*, Pittsburgh, or at www.aeaweb.org/articles.php?doi=10.12575/aer.98.2.1), (2008)

Stern, N., 'Key elements of a global deal on climate change' LSE, (2008b)

Sterner, T. and Persson, U. M., 'An Even Sterner Review: Introducing Relative Prices into the Discounting Debate', *Resources for the Future Discussion Paper*, (2007)

Sunstein, C. and Thaler, R., *Nudge: Improving Decisions About Health, Wealth and Happiness* (Yale University Press, New Haven, 2008)

Tamirisa, N., Jaumotte, F., Jones, B., Mills, P., Ramcharan, R., Scott, A., and Strand, J. , Chapter 4: 'Climate Change and the Global Economy', *World Economic Outlook*, (International Monetary Fund, Washington DC, April 2008)

Thompson, L. G., Yao, T., Mosley-Thompson, E., Davis, M. E., Henderson, K. and P.-N. Lin, 'A high-resolution millennial record of the South Asian Monsoon from Himalayan Ice Cores', *Science* 289, (2000)

Tol, R. S. J., 'Estimates of the Damage Costs of Climate Change, Part II: Dynamic Estimates', *Environmental and Resource Economics*, (2002)

Torras, M., 'The total value of Amazonian deforestation 1978–1993', *Ecological Economics 33* (2006)

United Nations Development Programme, *Human Development Report 2007/2008 Fighting Climate Change: Human Solidarity in a Divided World,* (Palgrave Macmillan, New York, 2007)

United Nations Food and Agricultural Organisation (FAO), 'Global Forest Resources Assessment 2005: Progress towards sustainable forest management', (FAO Corporate Document Repository, Rome, 2005)

Watson, R. T. and the Core Writing Team (eds), IPCC, *Climate Change 2001: Synthesis Report: A Contribution of Working Groups I, II and III to the Third Assessment Report of the Intergovernmental Panel on Climate Change* (CUP, Cambridge, 2008)

Weitzman, M. L., 'A Review of the Stern Review on the Economics of Climate Change', *Journal of Economic Literature*, (2007)

Weitzman, M. L., 'On Modeling and Interpreting the Economics of Catastrophic Climate Change' (unpublished , 2007b)

'World Development Indicators', (World Bank, Washington DC, 2008), (www.web.worldbank.org/WBSITE/EXTERNAL/DATA STATISTICS/0,,contentMDK:21725423~pagePK:64133150~piPK:641 33175~theSitePK:239419,00.html).

'World Oil Outlook to 2030', *Oil & Gas Journal*, (2008)

Worsnip, P., 'UN says credit crisis could enable "green growth"', Reuters, (www.alertnet.org/thenews/newsdesk/N10403551.htm), (10 October 2008)

Index